AF761354

NANOTECHNOLOGY SCIENCE AND TECHNOLOGY

CELLULOSE TO NANOCELLULOSES

PRODUCTION, PROPERTIES AND APPLICATIONS

Nanotechnology Science and Technology

Additional books and e-books in this series can be found on Nova's website under the Series tab.

NANOTECHNOLOGY SCIENCE AND TECHNOLOGY

CELLULOSE TO NANOCELLULOSES

PRODUCTION, PROPERTIES AND APPLICATIONS

TRI-DUNG NGO
EDITOR

nova
science publishers
New York

Copyright © 2019 by Nova Science Publishers, Inc.

All rights reserved. No part of this book may be reproduced, stored in a retrieval system or transmitted in any form or by any means: electronic, electrostatic, magnetic, tape, mechanical photocopying, recording or otherwise without the written permission of the Publisher.

We have partnered with Copyright Clearance Center to make it easy for you to obtain permissions to reuse content from this publication. Simply navigate to this publication's page on Nova's website and locate the "Get Permission" button below the title description. This button is linked directly to the title's permission page on copyright.com. Alternatively, you can visit copyright.com and search by title, ISBN, or ISSN.

For further questions about using the service on copyright.com, please contact:
Copyright Clearance Center
Phone: +1-(978) 750-8400 Fax: +1-(978) 750-4470 E-mail: info@copyright.com.

NOTICE TO THE READER

The Publisher has taken reasonable care in the preparation of this book, but makes no expressed or implied warranty of any kind and assumes no responsibility for any errors or omissions. No liability is assumed for incidental or consequential damages in connection with or arising out of information contained in this book. The Publisher shall not be liable for any special, consequential, or exemplary damages resulting, in whole or in part, from the readers' use of, or reliance upon, this material. Any parts of this book based on government reports are so indicated and copyright is claimed for those parts to the extent applicable to compilations of such works.

Independent verification should be sought for any data, advice or recommendations contained in this book. In addition, no responsibility is assumed by the Publisher for any injury and/or damage to persons or property arising from any methods, products, instructions, ideas or otherwise contained in this publication.

This publication is designed to provide accurate and authoritative information with regard to the subject matter covered herein. It is sold with the clear understanding that the Publisher is not engaged in rendering legal or any other professional services. If legal or any other expert assistance is required, the services of a competent person should be sought. FROM A DECLARATION OF PARTICIPANTS JOINTLY ADOPTED BY A COMMITTEE OF THE AMERICAN BAR ASSOCIATION AND A COMMITTEE OF PUBLISHERS.

Additional color graphics may be available in the e-book version of this book.

Library of Congress Cataloging-in-Publication Data

ISBN: 978-1-53615-057-5

Published by Nova Science Publishers, Inc. † New York

CONTENTS

PREFACE

The gradual depletion of petroleum resources, the lack of space for landfills, concerns over emissions during incineration, and environmental pollution caused due to accumulation of these non-destructible solid wastes has spurred efforts to develop high performance materials which are eco-friendly and sustainable. Cellulose constitutes the most abundant, renewable polymer resource. The materials have been considered to be the sustainable source of organic carbon in earth and the perfect equivalent to petroleum for the production of chemicals with low or net zero carbon emission. Cellulose based materials have been utilized by our society as engineering materials for thousands of years and their use continues today as verified by the enormity of the worldwide industries in forest products, construction, paper, textiles, etc. Keeping in mind the advantages of the bio-based materials, this book compiled in response to the growing interest of cellulose from micro- to nanoscale (nanocellulose). This book covers micro and nano technology related to cellulose and nanocellulose including chemistry, production, properties and applications of the cellulose based materials in various areas. From the view of sustainable development, the materials associated with cellulose bio-renewable sources are enormously being addressed. In addition to this, the book also summarizes the recent developments made in the area of advanced bio-nanomaterials, chemical functionalization of celluloses from micro-to nanoscale, theirs processing and successful utilization for selected applications. A number of critical issues and suggestions for future work are discussed, underscoring the roles of researchers for the efficient development of advanced bio-nanomaterials through value addition to enhance their use.

The book consists of 10 chapters, and each chapter can be regarded as a self-standing chapter. The chapters summarize recently developed research concerning biorenewable sources from cellulose to nanocellulose. Each chapter gives an overview of a particular material, its processing, and successful utilization for selected applications.

Chapter 1, "Cellulose: source, properties and applications" discusses and highlights cellulose, its diversity of products and the promising prospects for applications as this ancient material has found new scientific applications in various sectors nowadays.

Chapter 2, "Nanocellulose: from production to applications", highlights the roadmap from cellulose to its own different derivatives at nanoscale using agriculture- and forest-based products. Various techniques to produce nanocellulose are also covered in this chapter. This paves the way not only to diversify the renewable agriculture and forest product industry through nanotechnology but also to engineer and develop applications of novel nano-based biomaterials.

Chapter 3, "Synthetic methods to produce carboxylated cellulose nanocrystals", covers an overview of the various literature methods to synthesize carboxylated cellulose nanocrystals (cCNCs) and a description of some of their unique features. The common goal of these procedures is to develop a simple high yielding process that oxidizes the surface primary hydroxyl groups with high conversion rate, while leaving the secondary hydroxyl groups and the crystallinity intact.

Chapter 4, "Cellulose nanofiber, properties and applications", focuses on different isolation methods of cellulose nanofiber as well as detailed discussions on its properties such as mechanical properties, thermal degradation properties, and crystallinity index. In addition, several potential applications of cellulose nanofiber such as composite, smart materials, food packaging, coating, drug delivery and other are also going to be discussed.

Chapter 5, "Significance of the characterization activities in the production of cellulose nanomaterials", provides an overview on the significance of characterization activities as an essential part of the production of cellulose nanocrystals that helped optimize cellulose nanomaterials (CNMs) production processes, yields and quality, build capability, knowledge, and expertise in the CNMs characterization area. This chapter also outlines develop and co-develop methods for CNC characterization and establish protocols for quality control of CNMs.

Chapter 6, "Cellulose derived bionanocomposites", highlights processing strategies and common factors affecting the cellulose nanocrystals (CNC)/cellulose nanofibrils (CNF) based bionanocomposite preparation. It also covers the recent research on cellulose-reinforced bionanocomposites, where CNC or CNF is used as a nanophase in synthetic or bio-based polymers or their derivatives. The mechanical and thermal properties of the CNC and CNF derived bionanocomposites are discussed. The chapter also outlines the future perspective and current challenges related to CNC and CNF based bionancomposites.

Chapter 7, "Interfaces in aromatic polymer - nanocellulose composites", reviews on synthetic and processing techniques for improving interfacial properties between cellulose nanocrystals and aromatic polymers. Measurement techniques for examining the interface and associated properties dominated by the interface are discussed. The cahpter also reports results using a CNM modification proceess developed in laboratory to improve dispersion in polymer composites.

Chapter 8, "Barrier properties of nanoclellulose based films and coatings", provides a comprehensive overview of the current advances in mositure and oxygen barrier properites of nanocellulose based coatings and films. The chapter also covers the brief discription of molecule transport mechanism through films or coatings and factors affecting the barrier properties. The discussion includes the viewpoints on future research of CNC/CNF derived films and coatings to substitute synthetic materials with new cellulose bionanocomposites.

Chapter 9, "A review on mechanical behavior of cellulose and nanocellulose composites", attempts to provide a short summary on the mechanical properties and behavior of cellulose fibers, nanocellulose, and the cellulose and nanocellulose composites. Addtionally, the challenges associated with cellulose and nanocellulose based composites with and without surface modifications are discussed.

Chapter 10, "Stability and mobility of foam generated by gas-solvent/nanocellulose under reservoir condition", covers systematically evaluation the static stability and dynamic mobility of C_3H_8 foam stabilized by nanocellulose experimentally by using a pressure/volume/temperature (PVT) setup and a glass beadpack, respectively. Static foam stability test is conducted by recording the foam height as a function of time. Dynamic mobility test is conducted by recording the pressure drop across the porous glass beadpack.

As the editor of *Cellulose to Nanocellulose: Production, Properties and Applications*, I have attempted to compile, unify, and present the emerging research trends in materials based celluloses. I hope this book will contribute to the advancement of both science and technology in this exciting area. It is highly recommended for graduate students, engineers, researchers/faculty and professionals/practitioners in different fields such as chemistry, chemical engineering, civil engineering, forestry and agriculture, pulp and paper producers, and not limit to end users as paper and paperboard, paints and coatings, composites, natural textiles, oil and gas, cements and others.

I would like to thank all contributors, who provide excellent contributions and commitments on the chapters and discussions for this book. I also acknowledge Mrs. Marian Parslow, Mr. Brent Scorfield and Mr. Robert Semeniuk for helping in editing of the chapters. I wish to express my appreciation to my family for helping me to know which path to follow. I would like to extend my appreciation to the publishing staff at Nova Science Publishers for their support. Without them, this book will never be realized.

Tri-Dung (T.-D.) Ngo, PhD

ABBREVIATIONS

^{13}C NMR	^{13}C Solid State Nuclear Magnetic Resonance
ACC	Aqueous Counter Collision
AES	Auger Electron Spectroscopy
AGU	Anhydroglucopyranose Units
AFM	Atomic Force Microscopy
APS	Ammonium Persulfate
BEP	Bleached Eucalyptus Pulp
BET	Brunauer, Emmett and Teller, surface area analysis
CA	Cellulose Acetate
CAB	Cellulose Acetate Butyrate
CAGR	Compound Annual Growth Rate
CAP	Cellulose Acetate Phthalate
CF	Cellulose Microfibers
CM	Confocal Microscopy
CMC	Carboxymethyl Cellulose
CNC	Cellulose Nanocrystals
cCNCs	Carboxylated Cellulose Nanocrystals
CNF	Cellulose Nanofibers
CNF	Cellulose Nanofibrils
CNM	Cellulose Nanomaterials
CNT	Carbon Nanotube
CED	Cupriethylenediamine
CFP	Cotton based Filter Paper
CSA	Canadian-Standard Association
CTMP	Chemithermomechanical Pulping

DGEBA	Diglycidyl Ether of Bisphenol A
DLS	Dynamic Light Scattering
DO	Degree of Oxidation
DP	Dissolving Pulp
DP	Degree of Polymerization
DSC	Differential Scanning Calorimetry
EA	Elemental Analysis
EC	Ethyl Cellulose
EDS, EDX or XEDS	Energy-dispersive X-ray Spectroscopy
FC	Flow Cytometry
FE-SEM	Field Emission Scanning Electron Microscopy
FDA	Food and Drug Administration
FTIR	Fourier Transform Infrared Spectroscopy
FPLC	Fast Protein Liquid Chromatography
GO	Graphene Oxide
GW	Groundwood
HDPE	High Density Polyethylene
HMF	2-Hydroxymethylfurfural
HPC	Hydroxyethyl Cellulose, Hydoxypropyl Cellulose
HPMC	Hydroxylpropylmethyl Cellulose
HW	Hardwood
ICP-AES	Inductively Coupled Plasma Atomic Emission Spectroscopy
ICP-MS	Inductively Coupled Plasma Mass Spectrometry
ICP-OES	Inductively Coupled Plasma Optical Emission Spectrometry
IL	Ionic Liquid
IR	Infrared Spectroscopy
LDPE	Low Density Polyethylene
LVTEM	Low Voltage Transmission Electron Microscopy
MC	Methyl Cellulose
MCC	Microcrystalline Cellulose
MFC	Microfibrillated cellulose
NaClO	Sodium Hypochlorite
NaCMC	Anionic Ether Derivatives Like Sodium Carboxymethyl Cellulose
NCC	Nanocrystals Cellulose

NFC	Nanofibrillated Cellulose
NMR	Nuclear Magnetic Resonance
OLED	Organic Light Emitting Diode
PBAT	Poly(butylene adipate-co-terephthalate)
PE	Polyethylene
PEG	Polyethylene Glycol
PET	Polyethylene Terephthalate
PHA	Polyhydroxyalkanoates
PLA	Polylactic Acid or Polylactide
PP	Polypropylene
PS	Polystyrene
PVC	Polyvinyl Chloride
PVT	Pressure/Volume/Temperature
RGO	Reduced Graphene Oxide
RMP	Refiner Mechanical
RS	Raman Spectroscopy
s-CNC	Sulfated Cellulose Nanocrystals
SAG	Surfactant Mixtures
SDP	Spruce Dissolving Pulp
SEM	Scanning Electron Microscopy
SMC	Sheet Molding Compounding
SLS	Static Light Scattering
STEM	Transmission Mode SEM
t-CNC	Tunicate Nanocrystals
T_g	Glass Transition Temperature
TEM	Transmission Electron Microscopy
TEMPO	2,2,6,6-tetramethylpiperidine-1-oxyl radical
TiD	Initial Degradation Temperature
TGA	Thermal Gravimetric Analysis
TMP	Thermomechanical Pulping
TPE	Thermoplastic Elastomers
UV-Vis	UV-visible Spectroscopy
XPS	X-ray Photoelectron Spectroscopy
XRD	X-ray Diffraction
ZP	Zeta Potential

In: Cellulose to Nanocelluloses
Editor: Tri-Dung Ngo
ISBN: 978-1-53615-057-5
© 2019 Nova Science Publishers, Inc.

Chapter 1

CELLULOSE: SOURCE, PROPERTIES AND APPLICATIONS

Tri-Dung Ngo*
InnoTech Alberta Inc., Edmonton, Alberta, Canada
Edmonton, Alberta, Canada
University of Alberta, Faculty of Engineering,
Department of Civil and Environmental Engineering,
Edmonton, Alberta, Canada

ABSTRACT

Every plant is made up of cellulose, the constituent building block that comprises cell walls to make stems, branches and leaves strong. Cellulose, a linear polysaccharide polymer with many glucose monosaccharide units, and is recognized as the most abundant organic compound on earth. It has been used historically in paper-making and is today used in the production of many low-molecular weight organic products. Additionally, a number of useful derivatives are manufactured from cellulose. These regenerated celluloses are man-made, or synthetic and can be ultilized in a wide variety

* Corresponding Author Email: tridung.ngo@innotechalberta.ca.

of applications such as clothing textile, disposable medical devices and artificial membranes. With the growing global demand for environmentally-friendly, versatile, biodegradable, and cost-effective materials, cellulose is considered to be a good potential substrate with a market expected to witness significant growth. This chapter discusses and highlights cellulose and its uses in a diverse range of products and the promising prospects for applications.

Keywords: cellulose, cellulose derivatives, fiber

INTRODUCTION

Cellulose comprises at least one third of the vegetable material in the world and is the most abundant organic polymer on earth. Cellulose is a natural polymer most frequently found in the plant kingdom [1] and classified as a polysaccharide. It was discovered in 1838 by Anselme Payen, a French chemist who isolated it from plant matter and determined its chemical formula [2]. Cellulose comprises about 70% of wood and is subdivided into α-cellulose and hemicellulose. Pulp is a lignocellulosic fibrous material prepared by chemically or mechanically separating cellulose fibers from wood, fiber crops, or rags. As early as the second century, pulp for use in papermaking was produced by macerating mulberry bark. Many kinds of paper have been made exclusively from wood fibers with no other material additives in the formula. Large-scale wood paper production began with the development of mechanical pulping in Germany by Friedrich Gottlob Keller in the 1840s, and by the Canadian inventor Charles Fenerty in Nova Scotia. Roth discovered chemical pulping processes by using sulfurous acid to treat wood in 1857. Benjamin Tilghman invented the use of calcium bisulfite, $Ca(HSO_3)_2$, to pulp wood in 1867. Almost a decade later, the first commercial sulfite pulp mill was built in Sweden and by 1900; sulfite pulping had become the dominant means of producing wood pulp for paper making, surpassing the mechanical pulping methods [3, 4].

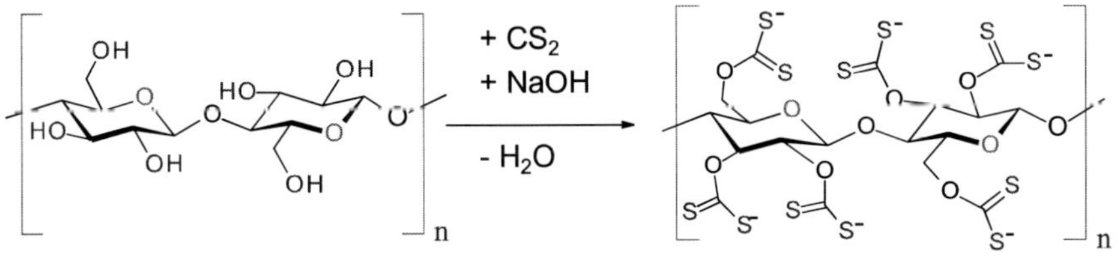

Figure 1. Treatment of cellulose with alkali and carbon disulfide.

In 1846 cellulose nitrate, which was commonly named nitrocellulose, was discovered by Schönbein. George Audemars first manufactured regenerated nitrocellulose fibers in 1855. Regenerated cellulose can be ultilized for a wide variety of applications such as oner of the first products created from regenerated cellulose, clothing textiles. The regenerated cellulose material is also used in the production of disposable medical devices as well as in the fabrication of artificial membranes. Though these fibers are soft and strong, resembling silk, they are also highly flammable, a significant drawback. Hilaire de Chardonnet perfected the process for production of nitrocellulose fibers but manufacturing of these fibers was relatively uneconomical [5]. Schweizer established the method for preparation of Schweizer's reagent, i.e., a cuprammonium hydroxide solution representing the first cellulose solvent in 1857. The crystalline structure of cellulose was first established by Carl von Nägeli in 1858 [6], and was later verified by X-ray crystallography. An organo-soluble cellulose acetate was synthesized by Schützenberger in 1865.

Other uses for cellulose were also emerging. It was successfully utilized to produce a thermoplastic polymer, celluloid, by Hyatt manufacturing company in 1870. Production of rayon, or "artificial silk," from cellulose began in the 1890s. During this period, Despeissis invented the cuprammonium process which uses a cuprammonium solution to solubilize cellulose, a method still used today for production of artificial silk. In 1891, it was discovered that treatment of cellulose with alkali and carbon disulfide (as shown in Figure 1) generated a soluble cellulose derivative known as viscose [3, 7]. The process was patented by the founders of the Viscose Development Company and remains the most widely used method for manufacturing regenerated cellulose products. Courtaulds purchased the patents in 1904, leading to significant growth of viscose fiber production. Cellophane, a thin, transparent sheet made of regenerated cellulose, was invented by Swiss chemist Jacques E. Brandenberger while employed by Blanchisserie et Teinturerie de Thaon. In 1912, he constructed a machine to manufacture the film which he named Cellophane. Its low permeability to air, oils, greases, bacteria, and water makes it useful for food packaging. Hermann Staudinger determined the polymer structure of cellulose in 1920. The introduction of solidstate CP/MAS 13C-NMR in the early 1980s provided new insight into the structure of cellulose [8, 9]. Cellulose compound was first chemically synthesized without the use of any biologically derived enzymes in 1992, by Kobayashi and Shoda [10].

Although discovered as early as 1949, cellulose nanocrystals did not receive much attention until 1992, when their colloidal suspensions showed unexpected liquid crystalline properties. The immense strength and stiffness of cellulose crystallites soon spurred interest in their potential to reinforce polymers. In recent years, the research studies on polymeric materials from renewable natural sources have been growing markedly [11]. The interest in utilization of renewable materials such as cellulose and its derivatives is directly related to a requirement to reduce society's dependence on materials obtained from fossil sources [12, 13]. In addition, these renewable materials also exhibit properties such as biocompatibility, biodegradability, low toxicity, and low cost among others [14].

Cellulose Sources and Structure

Cellulose is an important structural component of the primary cell wall of green plants, many forms of algae and the oomycetes [15]. Cellulose constitutes the most abundant, renewable polymer resource available worldwide.

Cellulose Sources

Celluloses are broadly classified into two categories, namely, natural and man-made. The natural cellulose fibers include cotton, jute, and so on, whereas man-made cellulose fibers include viscose, lyocell, modal, and others. Both kinds of fibers are widely used in the manufacture of textile. The primary natural sources of cellulose is the existing lignocellulosic material found in forests. Wood is the most important source of cellulose from those forests. Softwood and hardwood are distinguished botanically in terms of their reproduction, not by their end use or appearance. All trees reproduce by producing seeds, but the seed structure varies. Normally, one would expect the term "hardwood" to mean that the wood is hard to the touch, heavy, or dense, with the reverse being true of "softwood" [16]. Figure 2 shows the images of softwood and hardwood. However, the terms hardwood and softwood really have nothing to do with the hardness or softness of the wood. Hardwood is not necessarily a harder or more dense material and softwood is not necessarily a softer material or less dense as compared to its counterpart. These names, although misleading, refer to the type of growth or tree vegetation that they are taken from. In general, softwood comes from coniferous/cone bearing evergreen trees with needles. Hardwood comes from deciduous trees that are leaf bearing and drop their leaves annually in winter. Hardwoods tend to be slower growing, and are therefore usually more dense. Softwoods are generally closed-cell structures while hardwoods are generally open-cell with pores and vessels [17]. The closed-cell construction of softwood allows it to resist the penetration of water better than an open cell hardwood. Since most of the deterioration of wood is caused by wetting and drying, it also makes sense that the less penetration the better. Therefore, softwood is best used for outdoor projects.

Table 1. Chemical composition of some cellulose-containing materials [30-40]

Source	Cellulose (%)	Hemicellulose (%)	Lignin (%)	Extract (%)
Abaca	56.0-63.0	20.0-25.0	7.0-9.0	3.0
Bagasse	40.0	30.0	20.0	10.0
Bamboo	26.0-43.0	30.0	21.0-31.0	
Coir	32.0-43.0	10.0-20.0	43.0-49.0	4.0
Corn cobs	45.0	35.0	15.0	5.0
Corn stalks	35.0	25.0	35.0	5.0
Cotton	95.0	2.0	1.0	0.4
Curaua	71.0-74.0	9.9-21.0	7.5-11.0	2.5-2.8
Flax (retted)	71.0	21.0	2.0	6.0
Flax (unretted)	63.0	12.0	3.0	13.0
Grass	25.0-40.0	25.0-50.0	10.0-30.0	-
Hardwood	43.0-47.0	25.0-35.0	16.0-24.0	2.0-8.0
Hemp	70.0	22.0	6.0	2.0
Henequen	78.0	4.0-8.0	13.0	4.0
Istle	73.0	4.0-8.0	17.0	2.0
Jute	71.0	14.0	13.0	2.0
Kenaf	36.0	21.0	18.0	2.0
Oil Palm	65.0	29.0	-	
Pineapple	70.0-82.0	-	5.0-12.0	-
Ramie	76.0	17.0	1.0	6.0
Rice Husks	28.0-36.0	12.0-30.0	15.0-20.0	-
Rice Straw	29.0-35.0	23.0-26.0	17.0-19.0	-
Sisal	73.0	14.0	11.0	2.0
Sunn	80.0	10.0	6.0	3.0
Softwood	40.0-44.0	25.0-29.0	25.0-31.0	1.0-5.0
Wheat straw	30.0	50.0	15.0	5.0

Cellulose is surrounded by a matrix of hemicellulose and lignin in the secondary plant cell walls, which accounts for most of the carbohydrates in plant biomass; these three components are known collectively as lignocellulose [24]. Cellulose provides structural support and tensile strength, enabling plants to withstand stressors like weather and wind. Cellulose can be derived from other sources such as agricultural residues, water plants, grasses, and other plant substances, annual plants, microbes, and animals. These include seed fiber (cotton), bast fibers (flax, hemp, jute, ramie) and grasses (bagasse, bamboo). The cellulose content of cotton fiber is 90%, wood is 40-50%, and dried hemp is approximately 57% [25-27]. Table 1 shows chemical composition of some cellulose-containing materials. It has been estimated that by photosynthesis, 10^{11}-10^{12} tons of cellulose are synthesized annually in a rather pure form. Some species of bacteria secrete it to form biofilms [28]. Bacteria and algae also produce cellulose that is devoid of hemicellulose and lignin. Various cellulosic substrates originating from plants, bacteria and algae are used as model substrates in different kinds of studies [29].

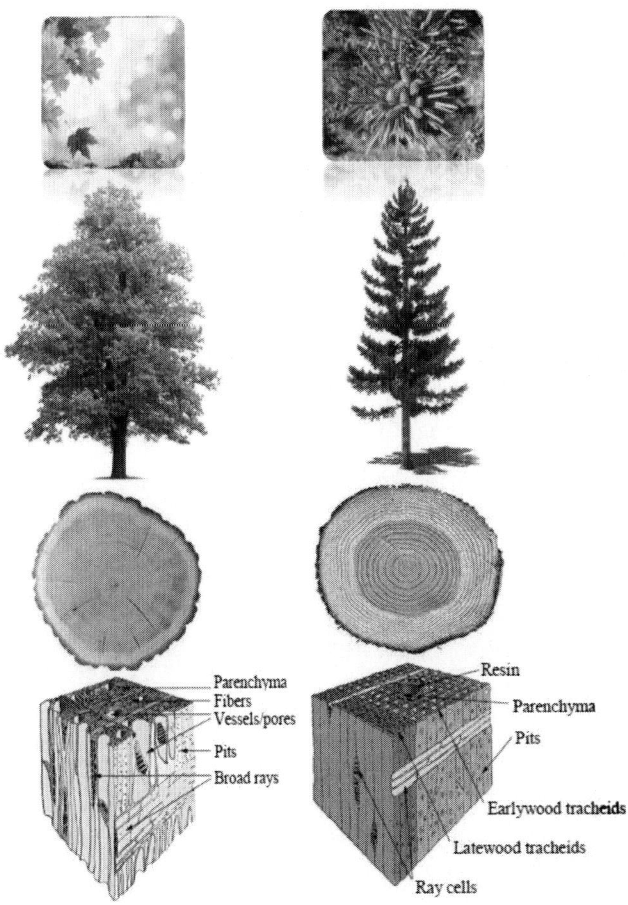

Figure 2. Hardwood and softwood [18-23].

Regenerated celluloses are man-made or synthetic, created from natural polymers. All forms of cellulose degrade before they melt, but natural cellulose can be regenerated by the xanthate process to manufacture cellulose fibers. The first man-made fibers which were developed and produced used polymers of natural origin, more precisely, cellulose, which is a raw material available in large quantities in the vegetable world. Viscose, triacetate, and acetate are the most common variants of rayon, which are man-made celluloses used extensively across end-use industries.

The chemosynthesis of functionalized cellulose by ring-opening polymerization of 3,6-di-O-benzyl-α-D-glucopyranose 1,2,4-orthopivalate, or by stepwise reactions of selectively protected β-D-glucose as, e.g., 1-allyl-2,6-di-O-acetyl-3-benzyl-4-O-(p-methoxybenzyl)-β-D-glucopyranoside [41, 42] has been experimentally realized. The nonbiosynthesis with controlled molecular weight preparation of cellulose was described involving an enzymic polymerization by Kobayashi in 1991 [43].

The cellulose products like methyl-, ethyl-, hydroxyalkyl ethers, cellulose acetate, as well as cellulose products with combinations of various functional groups, e.g., ethylhydroxyethyl cellulose, hydroxypropylmethyl cellulose, cellulose acetopropionates and cellulose acetobutyrates are still important in different applications, even many decades after their discovery. In additon, ionic cellulose derivatives have been recognized for a long time. Carboxymethyl cellulose, up to now, is the most important ionic cellulose ether which was first prepared in 1918 and produced commercially in the early 1920's in Germany [44]. These are produced in large quantities for diversified applications.

Cellulose Structures

Cellulose is an organic compound with the formula $(C_6H_{10}O_5)_n$. It is a polydisperse linear homopolymer, polysaccharide consisting of a linear chain of several hundred to many thousands of β(1→4) linked D-glucose units [45, 46]. This linkage motif contrasts with that for α(1→4)-glycosidic bonds present in starch and glycogen. The cellobiose as the repeating building blocks is composed of two adjacent anhydro-glucose rings connected via a β-1,4 glycosidic linkage as illustrated in Figure 3. Cellulose is similar in form to complex carbohydrates like starch and glycogen. These polysaccharides are also made from multiple subunits of glucose. The number of glucose units in cellulose varies from 7,000-10,000, however, the molecular structures of the polysaccharides are different depending on how the glucose molecules are linked together. The acetal linkage is beta which makes it different from starch. This peculiar difference in acetal linkages results in a major difference in digestibility in humans. Cellulose is a straight chain polymer with each cellulose molecule long and rod-like. This differs from starch, which is a coiled molecule. Table 2 shows the comparision between cellulose, glycogen and starch. Due to the results of these differences in structure, cellulose cannot be broken down into its glucose subunits by any enzymes produced from animals as is starch and other carbohydrates.

As a consequence, the hydroxyl groups are positioned in the ring plane (equatorial), while the hydrogen atoms are in the vertical position (axial). The polymer contains free hydroxyl groups at the C-2, C-3, and C-6 atoms. The degree of polymerization (DP) of cellulose, and consequently its molecular weight, depends on the cellulosic source, the processes and further treatments used in obtaining this biopolymer [47, 48]. The DP of cellulose can vary between 10,000 and 15,000 [49, 50]. Generally, the DP of cellulose profoundly influences the mechanical, solution, biological, and physiological properties of cellulose [51].

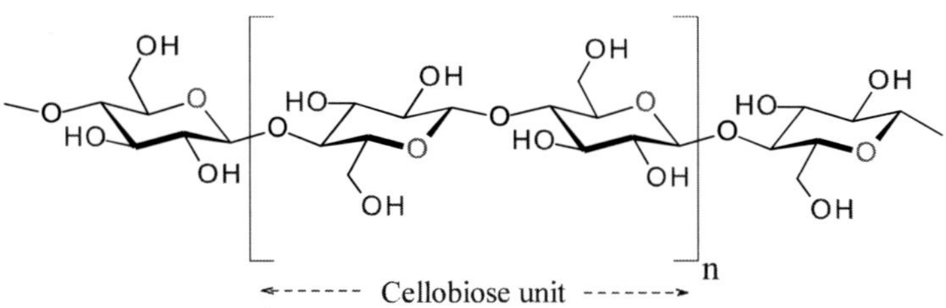

Figure 3. Schematic diagram of partial molecular structure of cellulose chain with numbering for carbon atoms and n= number of cellobiose repeating unit.

The structure of cellulose can be determined on the macro-, micro- and nano-scale, depending on the technique employed. Although researchers have devoted significant effort to investigating the structure of cellulose over several decades, this is not yet understood in sufficient detail. Terminal groups at either end of the cellulose molecule are quite different from each other. The C-1 OH at one end of the molecule is an aldehyde group with reducing activity. Aldehyde groups form a pyranose ring through an intramolecular hemiacetal form. In contrast, the C-4 OH on the other end of the chain is an alcoholborne OH constituent and thus is called the non-reducing end. It has been known from the infrared spectroscopy (IR), X-ray crystallography and nuclear magnetic resonance (NMR) investigations, that the repeated anhydroglucopyranose units (AGU) ring exists in the pyranose ring form and that this adopts the 4C1-chair formation which constitutes the lowest energy conformation for D-glucopyranose [52-56].

Table 2. Comparison between cellulose, glycogen and starch [57, 58]

Characters	**Cellulose**	**Glycogen**	**Starch**
Occurrence	in plants 	in animals and fungi 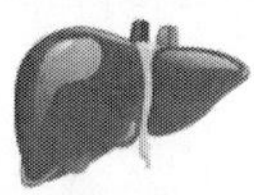	in plants 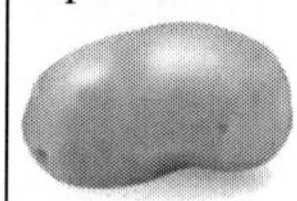
Function	structural support in plant cell walls	carbohydrate energy store	carbohydrate energy store
General form	fibers	small granules	grains
Monomer	β-glucose	α-glucose	α-glucose
Type of bond between monomers	1,4 glycosidic bond	1,4 and 1,6 glycosidic bonds	1,4 glycosidic bond (amylose) + 1,4 and 1,6 glycosidic bond (amylopectin)
Nature of chain	straight, long unbranched chains form H-bonds, with adjacent chains 	short many branched chains, some coiling 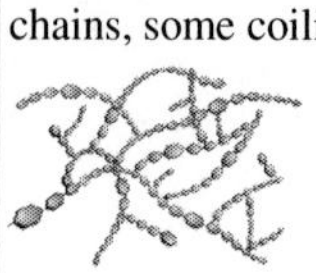	amylose is coiled unbranched amylopectin is long branched chains, some coiling 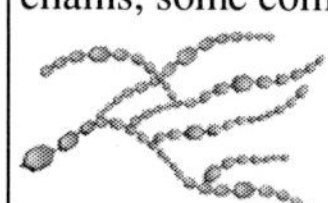

Acid degradation of cellulose was developed by Battista leading to the commercialization of microcrystalline cellulose (MCC). Most grades of MCC are made by hydrolysis of cellulose using HCl as the hydrolysis agent. Hydrolysis of MCC by H_2SO_4 may result in the formation of sulphated cellulose [59]. In recent years, cellulose has been utilized as a source for making nanocelluloses. The research on nanosized cellulose has experienced significant growth because of the unique combination of their properties including outstanding mechanical properties, rich surface chemistry, nontoxicity, biocompatibility, anisotropic shape, tailorable surface chemistry, interesting optical properties and availability from renewable sources. In addition, increased concern about the environment, demanded for sustainability in products is driven by consumers and market push for greater sustainability in the use of products and services, making the development and use of products from renewable resources one of the central priorities of the coming decades [60, 61].

CELLULOSE AND CELLULOSE DERIVATIVES PROPERTIES

Cellulose is odorless, tasteless, insoluble in water and most organic solvents, chiral, biodegradable and is hydrophilic with the contact angle of 20-30 degrees [62]. Cellulose has been shown to melt at 467°C [63]. It can be broken down chemically into its glucose units by treating it with concentrated mineral acids at high temperature [64]. One of the most specific characteristics of cellulose is that each of its monomers bears three hydroxyl groups. These hydroxyl groups and their hydrogen bonding ability play a major role in directing crystalline packing and in governing important physical properties of these highly cohesive materials. The ordered hydrogen bond systems on cellulose form various types of supramolecular semi-crystalline structures based on the OH groups and the oxygen atoms of both the pyranose ring and the glycosidic bond. It is the ability of these chains to hydrogen-bond together into fibers (microfibrils) that gives cellulose its unique properties of mechanical strength and chemical stability. Cellulose chains bond to each other forming elementary fibrils through hydrogen bonding. The elementary fibrils gather into microfibrils that are 5-50nm in diameter and several microns in length. Their structures contain disordered or amorphous-like and highly ordered or crystalline regions. The microfibrils of cellulose are extremely tough and inflexible due to the presence of hydrogen bonds. In fact, when describing the structure of cellulose microfibrils, chemists call their arrangement “crystalline,” meaning that the microfibrils have crystal-like properties. These cellulose fibrils are the main reinforcement phase for the subsequent cellular structured plant tissues and cellulose nanoparticles can be extracted from plant cell walls by chemical-mechanical processes.

Table 3. Properties of celluloses as compared to serveral biomass fibers and commonly used synthetic fibers [50, 68-76]

Fiber	Density (g/cm^3)	Elongation (%)	Tensile strength (MPa)	Young’s modulus (GPa)
Abaca	1.5	3.0-10.0	400.0	12.0
Bagasse	1.25	1.1	20.0-290.0	19.7-27.1
Bamboo	0.6-1.1	-	140.0-230.0	11.0-17.0
Banana	1.3-1.4	2.0-7.0	54.0-789.0	3.4-32.0
Coconut	1.4-3.8	-	120.0-200.0	19.0- 26.0
Coir	1.1	15.0-40.0	131.0-175.0	4.0-6.0
Cotton	1.5-1.6	3.0-10.0	287.0-597.0	5.5-12.6
Curaua	1.4	3.7-4.3	500.0-1150.0	11.8
Flax	1.4-1.5	1.2-3.2	345.0-1500.0	27.6-80.0
Hemp	1.48	1.6	550.0-900.0	70.0
Henequen	1.4	3.0-4.7	430.0-580.0	-
Jute	1.3-1.46	1.5-1.8	393.0-800.0	10.0-30.0
Kenaf	1.2-1.45	1.6	240.0-930.0	14.0-53.0
Nettle	1.51	2.1-2.5	560.0-1600.0	24.5-87.0
Oil Palm	0.7-1.55	25.0	248.0	3.2
Palf	1.4	3.0	170.0-635.0	6.2-24.6
Pineapple	1.5	1.6	413.0-1627.0	34.5-82.5
Ramie	1.5-1.56	1.0-3.0	400.0-1000.0	27.0-128.0
Sisal	1.33-1.5	2.0-14	400.0-700.0	9.0-38.0
Softwood kraft	1.5	-	1000.0	40.0
Rayon	1.6	1.25	500.0	40.0
Viscose (cord)	-	11.4	593.0	11.0
CNC	1.6	-	7500.0	120.0-220.0

Fiber	Density (g/cm^3)	Elongation (%)	Tensile strength (MPa)	Young's modulus (GPa)
CNF	1.5-1.6	-	2000.0	150.0
Carbon	1.4	1.4-1.8	1500.0-5500.0	230.0-240.0
Carbon nanotube, single-walled	-	16.0	13000.0-53000.0	1000.0-5000.0
Carbon nanotube, multi-walled	-	-	63000.0-150000.0	200.0-900.0
Kevlar	1.4	3.3-3.7	3000.0-3150.0	63.0-67.0
E-glass	2.5	2.5-3.0	2000.0-3500.0	70.0
S-glass	2.5	2.8	4570.0	86.0
Copper	8.9	-	200.0-480.0	116.0
Steel wire	7.8	-	4100.0	210.0
Nylon-6	1.13	90.0	45.0-90.0	2.0-4.0
Polyethylene HDPE (high density)	0.94-0.97	500.0	15.0	0.8
Polyethylene, LDPE (low density)	0.92	500.0	-	0.11-0.45
Polyethylene Terephthalate, PET	1.40	125.0	55.0	2.0-2.7
Polystyrene, PS	1.05	7.0	30.0-100.0	3.0-3.5
Polypropylene, PP	0.90	100.0	28.0-36.0	1.5-2.0

Cellulose is regarded as a semi-flexible biopolymer. Both crystalline and amorphous regions exist on the celluloses. The multiple hydroxyl groups on the glucose from one chain form hydrogen bonds with oxygen atoms on the same or on a neighboring chain, holding the chains firmly together side-by-side and forming microfibrils with high tensile strength. This confers tensile strength in cell walls, where cellulose microfibrils are meshed into a polysaccharide matrix. The relative stiffness and rigidity of the cellulose molecule is mainly due to the intramolecular hydrogen bonding. This property is reflected in its high viscosity in solution, a high tendency to crystallise, and its ability to form fibrillar strands.

In cellulose, intramolecular bonds occur between hydroxyl groups of adjacent cellulose molecules [65]. Van der Waals forces, in addition to these intermolecular hydrogen bonds are responsible for the occurrence of parallel stacking of multiple cellulose chains forming elementary fibrils that further aggregate into larger microfibrils [50]. Cellulose is insoluble in common organic solvents and in water. This is due to the fact that the hydroxyl groups are responsible for the extensive hydrogen bonding network forming both, intra- and intermolecular hydrogen bonding. In order to dissolve cellulose, the prevailing hydrogen bonding network must be broken. Cellulose is soluble in Schweizer's reagent, cupriethylenediamine (CED), cadmiumethylenediamine (Cadoxen), N-methylmorpholine N-oxide, and lithium chloride/dimethylacetamide [66]. This is used in the production of regenerated celluloses (such as cellophane and viscose) from dissolving pulp. Cellulose is also soluble in many kinds of ionic liquids [67]. Cellulose is also produced in a highly hydrated form by using bacteria (for example, acetobacter xylinum). The chemical character and reactivity of cellulose is determined by the presence of three equatorially positioned OH groups in the AGU, one primary and two secondary groups.

Information about mechanical properties of the materials are shown in Table 3 and can be compared to properties of commonly used fibers such as glass fiber and aramid fiber. However, the mechanical and physical properties of the natural cellulose fibers are mainly influenced by their composition: cellulose, hemicellulose and lignin. Hemicellulose is one of the fiber components responsible for the initial thermal degradation behavior and is also associated with the moisture content. Therefore, fibers containing high hemicellulose content should absorb more moisture and degrade at a lower temperature. The cellulose materials offer high specific proporties and can be compared to kevlar fiber in stiffness, making them excellent materials for reinforcement of composites and nanocomposites. Table 3 shows the properties of celluloses as compared to several biomass fibers and commonly used synthetic fibers.

Cellulose and Cellulose Derivatives Applications

The global cellulose fiber market is anticipated to reach USD 48.37 billion by 2025, growing at a compound annual growth rate (CAGR) of 9.1% from 2016 to 2025. The synthetic cellulose fiber segment dominated the market with 42% of the overall volume share in 2015 with textile as the prominent segment, alone accounting for around 44% of the global revenue share. The major application of fibers in the textile industry is to reinforce composites and filter chemicals [77]. Therefore, fiber produced from wood pulp is becoming an important element for the textile industry.

Cellulose Used in Textile, Pulp and Paper

Our society has ultilized natural cellulose based materials such as cotton, flax, hemp, wood, etc. in materials for thousands of years. A textile is a flexible material made of primarily polymeric either natural or synthetic fibers. The fiber can either be arranged directly into a nonwoven or into a yarn to form fabric. Historical textiles have been shown to have been made from cellulose fibers. These materials serve to create a special link between the natural environment and the social environment that underlies our lives, from the everyday textile to ceremonial to patriotic.

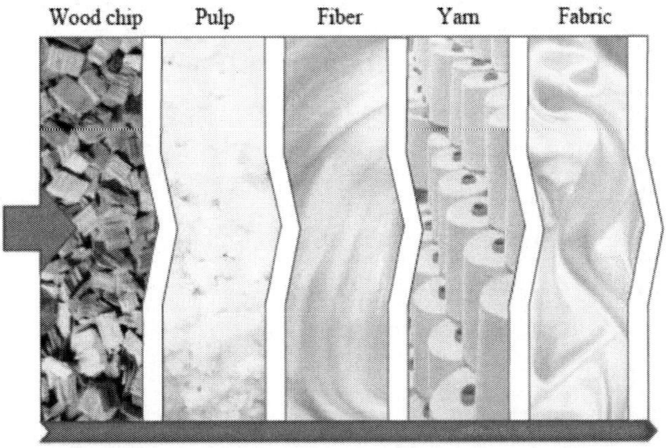

Figure 4. Turning wood into fibers that can be woven to make fabrics.

Cotton is the oldest and most common natural fiber, native to the tropical and subtropical regions of North America, central and South America, Egypt and India. The cotton fiber grows around the seeds of the plant and consists mainly of cellulose. The production of textile materials has undergone dramatic changes in the last century. Prior to the industrial revolution in the 19th century, natural materials, e.g., animal furs, cotton and silk had been used for thousands of years. In the first decades of the 20th century, however, cotton took over as the most popular raw material, accounting for more than 70% of all textile raw material production in the world [78]. These cellulose fibers are spun into threads for the textile industry. Cotton is a highly durable fiber with the ability to absorb moisture. Cotton, flax, hemp, jute, ramie, and other cellulosic fiber plants are all sources capable of producing textiles and fabrics that can be used to create knitted, woven, or nonwoven cloth material or fiber and yarn intended for fabric production. Regenerated cellulose fibers are made from cellulose extracted from wood. Regenerated cellulose fibers were first developed in France in the late 1800s. In the 1920s the textile industry started calling these fibers "rayon." Other forms of rayon in use today include lyocell, modal and viscose. The first commercial plant for viscose production was built in France in 1891 [79]. The production of textile materials is continuously growing with increasing global collective. Cellulose demand is expected to grow significantly over the forecast period owing to its increasing application in beauty, clothes, food, home furnishing and wash and wipe products.

Turning wood into fibers that can be woven into fabrics is a chemically intensive process. The starting point is wood chips from commonly used woods spruce, pine and beech or bamboo pith. The production of synthetic or man-made fibers includes the industrial processing of wood pulp derived from botanical sources. However, the manufacturing of cellulose fibers entails mechanical and chemical processing of wood pulp [80]. The materials are digested and produce a cellulose pulp whichis then further chemically processed, pressed, dried and shredded to produce a dry cellulose compound. This is dissolved to create a solution and then forced through a spinneret, a device that resembles a shower head, to produce strands. These strands are then drawn and stretched to produce long straight fibers. Yarn is generally defined as a continuous strand of fibers spun together as a group which can then be used to make fabrics as shown in Figure 4.

Apart from water, wood and other plant fibers used to make pulp contain three main components. The first is cellulose fibers which are desirable for papermaking. The second component is lignin, which is a three-dimensional polymer that binds the cellulose fibers together. The third component is hemicelluloses or shorter, branched-carbohydrate polymers. The pulp and paper industry comprises companies that use wood or other plant fiber as raw materials and produce pulp, paper, paperboard and other cellulose-based products. A pulp mill converts wood chips or other plant fiber sources into a thick fiberboard which can be shipped to a paper mill for further processing. The aim of pulping is to break down the bulk structure of the fiber source, be it chips, stems or other plant parts, into the constituent fibers. The finished product may be either bleached or non-bleached, depending on the customer requirements. Pulp can be manufactured using mechanical, semi-chemical or fully chemical-based methods such as kraft and sulfite processes [81, 82].

The various mechanical pulping methods, such as groundwood (GW) and refiner mechanical (RMP) pulping, physically separate the cellulose fibers one from another. Strength is impaired because the fibers may be cut. The lignin adheres to the fibers that can cause photo-yellowing of paper. In addition, the amount of energy subsequently required by the mechanical treatment is high. Chemical pulping degrades the lignin and hemicellulose into small, water-soluble molecules which can be washed away from the cellulose fibers without depolymerizing the cellulose fibers. Chemically depolymerizing the cellulose weakens the fibers during the pulping process. Alternatively, hybrid pulping methods use a combination of chemical and thermal treatment to begin an abbreviated chemical pulping process, followed immediately by a mechanical treatment to separate the fibers. These hybrid methods include thermo-mechanical pulping (TMP), and chemi-thermomechanical pulping (CTMP). The chemical and thermal treatments reduce the amount of energy subsequently required by the mechanical treatment and also reduce the amount of strength loss suffered by the fibers [81, 82].

The world paper output increased +0.8%y/y to 410 million tons in 2017. Asia accounts for the bulk of the market with nearly half of the global output (and consumption) of paper and board [83]. The global pulp and paper chemicals market is expected to register significant growth during the forecast period from 2018 to 2023. This is mainly driven by the growing pulp and paper industry in emerging economies and increased recycling of paper. In the United States, about 70% of paper mills rely on recycled paper fiber to produce their products at less cost and with reduced energy requirements [84].

Cellulose Used in Composites

Ecological concerns have resulted in a renewed interest in environmentally-friendly composites using sustainable, biodegradable andrenewable resources. Composite materials are a class of material usually made by the combination of a matrix and some kind of reinforcement such as fiber. The combination of the ductility of the matrix with the high strength of the fiber creates a stronger material. As an example, cellulose in trees makes the trunks strong enough to hold up the tallest trees. Recently, cellulose fibers began to be used as a fiber-reinforcement material, especially in conjunction with polymers in biocomposites and fiber-reinforced plastic. Cellulose fiber reinforced composites have been used in numerous applications including automotive, building and construction industries, bicycle frames, sports, decking, panels, and window frames. The cellulose has an interesting dimension, which is the relationship between length and diameter of the fibers. This is a determining factor in the transfer of efforts to the matrix. Another interesting point is the irregular cross section of plant fibers as well as their appearance fibrillated, which exerts a positive influence on the anchoring of the fibers with the fragile matrix. The high volume percentage of permeable voids contributes to reduced weight, increased acoustic absorption and low thermal conductivity from the obtained components [74, 76].

Natural cellulose is renewable, being derived from forest and agricultural products; therefore, the composite materials would biodegrade after their service life without additional harm to the environment. Furthermore, natural fibers are known for their good specific properties that can be utilized to reinforce the matrix. Greater use of cellulose fiber from wood in producing composites also could be a boost to the paper industry by providing an important new use for wood pulp. Depending on the target applications, different sources of celluloses are utilized. These sources include wood flour, wood fibers, cellulose fibers, bast fibers from bamboo, cork, flax, hemp, jute and kenaf, or fibers from sunflower seed shells. The matrix are common petrochemical polymers polyethylene (PE), polypropylene (PP), thermoplastic elastomers (TPE) and polyvinyl chloride (PVC), or biopolymers such as bio-PE, poly(lactic acid) or polylactic acid or polylactide (PLA), polybutylene succinate (PBS), poly(butylene adipate-co-terephthalate) (PBAT) or polyhydroxyalkanoates (PHA) [72-76]. In addition, the biocomposites market is driven by government regulations regarding the increased use of environmentally friendly products and safety as well as recyclability of biocomposites as compared to glass and carbon fiber composites. Also, technological advancement in the mass production of biocomposites is one of the factors driving the biocomposites market globally, valued at USD 16.46 billion in 2016 andprojected to reach USD 36.76 billion by 2022 at a CAGR of 14.44% from 2017 to 2022 [85]. The incorporation of bioresources such as natural cellulose fiber into composite materials can reduce further dependency on petroleum reserves. Wood-plastic composites are materials made of fine wood fibers or wood flour mixed with thermoplastics. The applications of wood-plastic composites market is used extensively in areas like building and construction products, decking, molding and siding, fencing, automotive components, industrial and consumer goods, as well as many others. The global wood-plastic composites market is anticipated to grow from USD 3.65 billion in 2017 to $10.75 billion by 2026, at a CAGR of 12.79% between the forecast period of 2018-2026 [86]. The renewable resource based bio-materials are currently being developed but need further research to overcome their performance limitations. At the present level of technology, non-wood fibers like flax, hemp, kenaf, and sisal have attained commercial success in designing bio-composites. Another type of fiber i.e., grass fibers is gaining attention of scientists as a reinforcing fiber for composite applications.

Cellulose Used in Food

Cellulose is a carbohydrate or polysaccharide and has minimal environmental, health and safety risks. Cellulose can be considered a "green" material for several applications. Various forms of cellulose have been approved by the Food and Drug Administration (FDA) for use in many consumer products in the food and medical industries. With a rising awareness of fiber intake in the food industry, cellulose has become one of the most popular additives for this area. It can be found on ingredient lists under a variety of names, depending on which form is used. Cellulose has the same molecular structure regardless of the source as wood pulp, cotton, or other vegetable matter, however, depending on how cellulose molecules are bonded together and whether or not they are hydrated creates different "forms" [87, 88].

The human body is unable to digest cellulose because the appropriate enzymes to break down the beta acetal linkages are lacking. Undigestible cellulose is the fiber which aids in the smooth working of the intestinal tract. Adding cellulose to food allows an increase in bulk and fiber content without a major impact on flavo and because it is indigestible to humans, it has no caloric value. For this reason, cellulose has become a popular bulking agent in diet foods. Consumers who eat foods with high cellulose content feel full physically and psychologically without having consumed many calories. Cellulose binds and mixes easily with water and is often added to increase the fiber content of drinks and other liquid items when the gritty texture of regular fiber supplements would be undesirable [87].

Cellulose gel acts similarly to an emulsion, suspending ingredients within a solution and preventing water from separating out. The gelling action of cellulose when combined with water provides both thickening and stabilizing qualities in the food to which it is added. Cellulose gum or cellulose gel, which are hydrated forms of cellulose, are often used in sauces or other wet items like ice cream and frozen yogurt for both the thickening and emulsifying action. Cellulose may also be found on ingredient lists under the names carboxymethylcellulose, microcrystalline cellulose, or MCC. The thickening behavior of cellulose also allows for more air to be whipped into products like ice cream or whipped topping. Cellulose allows for the production of thick and creamy food items without the use of as much fat. It's ability to absorb moisture and coat ingredients in a fine powder makes it widely used in food products and is the choice for use in anti-caking applications. Shredded and grated cheeses, spice mixes, and powdered drink mixes are just a few of the many food items that take advantage of cellulose as an anti-caking agent [88].

Some animals can digest cellulose. When an herbivore like a cow consumes a plant, like grass, bacteria present in their intestinal tract are able to break down the cellulose into smaller parts that the cow's own digestive system can absorb.

Cellulose Used in Pharmaceutical Industries

Rising applications of cellulose in pharmaceuticals is further anticipated and leading to a surge in its market growth in near future. Many consumer products in the medical industry have utilized cellulose in things such as, but not limited to, artificial tears, cellophane, chronic wound treatment, as a scaffolding structure for tissue engineering, artificial blood vessels, cartilage, and filler material within drug tablets. With its beneficial properties such as zero toxicity, hygroscopicity, chemical inactivity and reversible adsorbency, MCC has been considered and used in the pharmaceutical industry as a tablet excipient [89]. Table 4 shows the properties and utilization of cellulose materials related to pharmaceutical applications. Adhesive polymers are synthetic, semi synthetic or natural macromolecules with capablility of attaching to skin or mucosal surfaces. Bioadhesives are considered as novel drug delivery systems. Very different types of polymers have been used as bioadhesive polymers. The ability of polymers to take up water from mucus and the pH of it's target are important factors determining the adhesive power of polymers. Cellulose derivatives especially cellulose ethers are widely used in bioadhesives. They are used in various types of formulations such as buccal, ocular, vaginal, nasal and transdermal, alone or in combination with other polymers. Cellulose ethers recently used in bioadhesives include nonionic cellulose ethers such as carboxymethyl cellulose (CMC) or hydroxylpropylmethyl cellulose (HPMC), ethyl cellulose (EC), hydroxyethyl cellulose, hydoxypropyl cellulose (HPC), methyl cellulose (MC), and anionic ether derivatives like sodium carboxymethyl cellulose (NaCMC) [90-99].

The cellulose ethers are generally hydrophilic and convert to hydrogel after exposure to water. Both soluble and insoluble cellulose ethers can absorb water and form a gel. After exposing these coated dosage forms with water, the coated polymers form a hyrogel and gradually dissolve in water until they disappear, but the insoluble cellulose ether coatings remain as a viscouse gel around tablets, and drug release is performed by diffusion of drug molecules within this layer. These two types of dosage forms are called dissolution-controlled and diffusion-controlled drug delivery systems, respectively. Despite cellulose ethers, the cellulose esters are generally water insoluble or water soluble in a distinct pH range. These polymers like cellulose acetate (CA), cellulose acetate butyrate (CAB) and cellulose acetate phthalate (CAP) do not form gel in the presence of water and are widely used for preparing pH-sensitive and semi-permeable microporous membranes. These membranes are employed for a wide variety of controlled release coatings of pharmaceuticals, especially in enteric or osmotic drug delivery devices. These polymers are beneficial in making different cellulosic membrane filters which can be applied to the pharmaceutical industry [87].

Extended release pharmaceuticals refer to dosage forms that allow a twofold or greater reduction in frequency of the drug administration in comparison with conventional dosage forms. EC is water insoluble material and often used in conjunction with water soluble polymers such as MC and HPMC in aqueous coating liquids [90, 91]. EC solutions in organic solvents such as ethanol can be thickened by HPMC or HPC [92, 93]. The oxidized cellulose fabric, such as gauze or cotton, resembles the parent substance and is insoluble in acids and water but soluble in dilute alkalis. In weakly alkaline solutions, it swells and becomes translucent and gelatinous. When wet with blood, it becomes slightly sticky and swells, forming a dark brown gelatinous mass. So, it is used in various surgical procedures, by direct application to the oozing surface except when used for homeostasis [94]. It is not recommended as a surface dressing for open wounds. Acetobacter xylinum is produced by a bacterium and strong enough to cope with blood pressure and works well with the body's own tissue. The material also carries a lower risk of developing blood clots than the synthetic materials currently in use [95, 96].

Table 4. Properties utilized and applications of cellulose materials [97-99]

Product applications	**Properties utilized**
Bulk laxatives	High water binding capacity, physiologically inert
Creams, lotions, ointments	Emulsion, film former, stabilizer, thickener
Denture adhesives	Long lasting adhesion, wet tack
Foamed products, shampoos	Foam stabilizer, high water binding capacity, suspending aid, thickener
Granualation aid, tablet binder	High strength binder
Jellies, salves	Film former, gelling agent, protective colloid, thickener
Suspensions, syrups	Suspending aid, thickener
Sustained release	Diffusion barrier, thickener
Tablet coating	Film former
Toothpaste	Binder, flavor stabilizer, suspending aid, thickener

Chemical functionalization and modification of cellulose materials to increase the utilization of cellulose by altering its properties have been reported. The functionalization and modification were acquired through the reactions of hydroxyl groups such as etherification, esterification, intermolecular crosslinking reactions, and macrocellulosic free radical reactions, particularly in the formation of graft cellulose copolymers [100]. Table 5 shows the chemical structures of some cellulose esters and ethers. As a naturally occurring material, cellulose may contain byproducts leading

Table 5. Chemical structures of cellulose esters and ethers [98, 101]

Cellulose derivatives	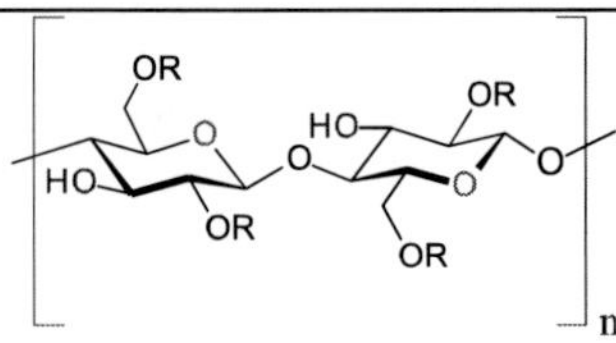
Cellulose esters	**R groups**
Acetate	H, 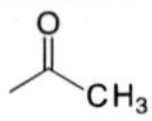
Acetate trimellitate	H, 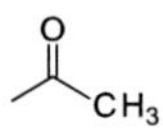, 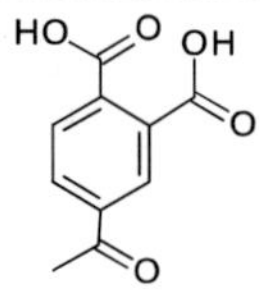
Acetate phthalate	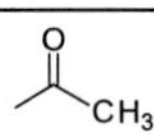, 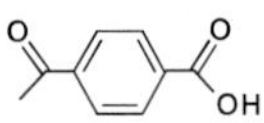
Hydroxypropylmethylphthalate	H, CH_3, $CH_2CH(OH)CH_3$, 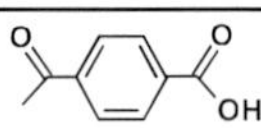, 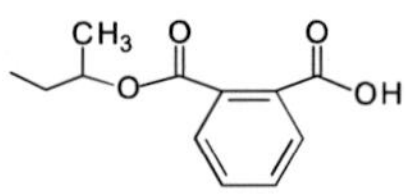
Hydroxypropylmethylphthalate acetate succinate	H, CH_3, $CH_2CH(OH)CH_3$, 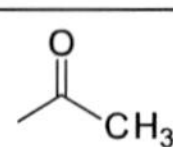, 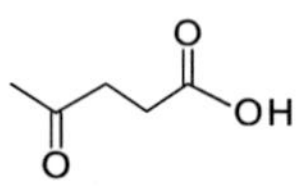
Nitrate	NO_2
Xanthate	CS_2Na
Carboxymethyl	H, CH_2COONa
Cyanoethyl	CH_2CH_2CN
Ethyl	H, CH_2CH_3
Hydroxyethyl	H, CH_2CH_2OH
Hydroxypropyl	H, $CH_2CH(OH)CH_3$
Hydroxypropylmethyl	H, CH_3, $CH_2CH(OH)CH_3$
Methyl	H, CH_3

to application problems and difficulties in chemical modification reactions. However, up-to-date methods of cellulose isolation and purification yield materials of high purity and variability. Table 6 presents some commercially marketed cellulose esters and ethers and their utilizations.

Table 6. Cellulose esters and ethers and their ultilizations [100, 102-104]

Cellulose derivatives	**DS range**	**Solubility**	**Product applications**
Cellulose esters			
Acetate	1.0-3.0	Acetone	Coatings, fibers, films, heat and rot resistant fabrics, membranes
Nitrate	1.5-3.0	MeOH, $PhNO_2$, ethanol-ether	Explosives, films, fibers, membranes
Xanthate	0.5-0.6	NaOH/water	Textiles
Carboxymethyl	0.5-1.2	Water	Coatings, detergents, fibers, food additives oil-well drilling muds, paper size, paints
Cyanoethyl	2.0	Organic solvents	Fabrics with heat and rot resistances, products with high dielectric constants
Ethyl	2.3-2.6	Organic solvents	Lacquers, pharmaceutical industry, plastics
Hydroxyethyl	Low DS	Water	Cosmetics, coating, films, paints
Hydroxypropyl	1.5-2.0	Water	Paints
Hydroxypropylmethyl	1.5-2.0	Water	Paints
Methyl	1.5-2.4	Hot water	Cosmetics, films, food additives, grease proof paper, textiles, tobacco industry

Cellulose ethers and esters are the two most important value-added cellulose derivatives with a wide range of applications. Most cellulose ethers are water soluble, exhibiting very different properties from the original cellulose. It is not surprising that the application of the ether derivatives are extremely versatile. Alkyl cellulose, carboxyalkylcellulose and hydroxyalkylcellulose are the major cellulose ethers derived through reactions on alkali cellulose. Cellulose nitrate and acetate are the predominant cellulose esters that find a market in plastics, lacquers, adhesives, films, fibers and explosives. The degree of substitution in these derivatives determines the properties and uses of the products. Viscose rayon and cellophane are obtained from cellulose xanthate as an intermediate. Improved strengths and chemical resistance can be achieved by graft copolymerization on cellulosic materials [100-104].

Others

Cellulose is among the oldest types of building insulation material. Many types of cellulosic materials have been used including newspaper, cardboard, cotton, straw, sawdust, hemp and corncob. Cellulose insulation is often made by hammer milling waste newspaper which is treated with chemicals, such as boric acid, to retard the spread of fire. Loose fill cellulose is next best after closed cell spray foam when sprayed into walls or into bib's (blown in blankets) stapled along the truss system because of it's ease of installation and lower cost compared to spray foam [105]. With the R-value per inch at about 3.7, celluloses (loose fill) are used for attic floor insulation and with R-value per inch about 3.2, celluloses (high density) are ulitized for the walls, enclosed cavities, and framing transitions.

Cellulose is also a source of energy for several types of microbes. Large amounts of cellulose is degraded to glucose by both bacterial and fungal microbes. The glucose derived from cellulose degradation is then used by other microbes as a carbon source to produce energy. Many fungi, specifically the members of Basidiomycetes, perform a critical ecological function by degrading cellulose present in decaying wood. Conversion of cellulose from energy crops into biofuels, such as cellulosic ethanol is under investigation as an alternative fuel source[106].

REMARKS

Cellulose has played a significant role in human culture, especially as paper. Today, this ancient material has found new scientific applications in various sectors. Cellulose, a unique and promising natural material, has gained much attention for its composites applications development. The ultilization of cellulose-based materials is continuing today, as verified by the enormity of the world wide industries using it, like forest products, paper, textiles, composites, etc. The attractive properties of cellulosic materials include biodegradability, biocompactibility, renewability, low density, high strength, good stiffness, and low thermal expansion. These qualities make them suitable for various applications. However, their consistency, uniformity, properties, functionality, and durability are still limited; therefore, the next generation of cellulose-based products will be required for certain engineering applications which cannot be achieved with traditional cellulosic materials.

REFERENCES

[1] Mccarthy, S. P., 2003. Biodegradable polymers. In: *Plastics and the environment*, Andrady, A. L., Ed., John Wiley and Sons, Inc., New York, 359-377. DOI: 10.1002/0471721557.ch9.

[2] Payen, A., 1838. Mémoire sur la composition du tissu propre des plantes et du ligneux (Memoir on the composition of the tissue of plants and of woody [material]), *Comptes Rendus* vol. 7, pp. 1052-1056.

[3] http://www.historyofinformation.com/detail.php?entryid=548.

[4] http://www.historyofinformation.com/detail.php?entryid=587.

[5] Volker, A., 2005. *Encyclopedia of polymer science and technology* (Wird aktualisiert. Ed.). Hoboken, N. J.: Wiley-Interscience. ISBN 9780471440260.

[6] Wilkie, J., 1961. Carl Nägeli and the fine structure of living matter. *Nature* 190:1145-1150. DOI:10.1038/1901145a0.

[7] Woodings, C., 2001. *Regenerated cellulose fibres.* [Boca Raton, FL: CRC Press; Cambridge, England: Woodhead Pub. ISBN 97818557 34593.

[8] VanderHart, D. L., Atalla, R., 1984. Studies of microstructure in native celluloses using solid-state carbon-13 NMR. *Macromolecules* 17(8), 1465-1472. DOI: 10.1021/ma00138a009.

[9] Kunze, J., Scheler, G., Schröter, B., Philipp, B., 1983. ^{13}C High resolution solid state NMR studies on cellulose samples of different physical structure. *Polymer Bulletin* 10(1), 56-62. DOI: 10.1007/ BF00263239.

[10] Young, R., 1986. Cellulose structure modification and hydrolysis. New York: Wiley. ISBN 0-471-82761-4.

[11] Petersson, L., Mathew, A. P., Oksman, K., 2009. Dispersion and properties of cellulose nanowhiskers and layered silicates in cellulose acetate butyrate nanocomposites. *Journal of Applied Polymer Science* 112, 2001-2009. DOI: 10.1002/app.29661.

[12] Feng, L., Zhou, Z., Dufresne, A., Huang, J., Wei, M., An, L., 2009. Structure and properties of new thermoforming bionanocomposites based on chitin whisker-graft-polycaprolactone. *Journal of Applied Polymer Science* 112, 2830-2837. DOI:10.1002/app.29731.

[13] Goetz, L., Mathew, A., Oksman, K., Gatenholm, P., Ragauskas, A. J., 2009. A novel nanocomposite film prepared from crosslinked cellulosic whiskers. *Carbohydrate Polymers* 75, 85-89. DOI: 10.1016/ j.carbpol.2008.06.017.

[14] Mecking, S., 2004. Nature or petrochemistry?-biologically degradable materials. angewandte chemie, *International Edition* 43, 1078-1085. DOI: 10.1002/anie.200301655.

[15] Klemm, D., Heublein, B., Fink, H.P., Bohn, A., 2005. Cellulose: fascinating biopolymer and sustainable raw material. *Angew. Chem. Int. Ed.* 44 (22), 3358-93. DOI:10.1002/anie.200460587. PMID 15861454.

[16] https://www.canadianwoodworking.com/get-more/hardwoods-softwoods.

[17] https://www.thoughtco.com/difference-between-hardwood-softwood-trees-1341883.

[18] http://www.noagergitavan.com/urun/sonbahar-yapraklari/.

[19] https://www.dreamstime.com/.

[20] https://www.barbequescience.com/wood-types/.

[21] https://hubpages.com/sports/Can-You-Use-Any-Type-Of-Pine-Needle-For-Pine-Needle-Tea-How-To-Make-Pine-Needle-Tea.

[22] https://free3d.com/3d-model/pine-tree-a-1944.html.

[23] https://www.ggspdt.com/42b-timber.html.

[24] Bidlack, J., Malone, M., Benson, R., 1992. Molecular structure and component integration of secondary cell walls in plants. *Proc. Okla. Acad. Sci.* 51-56.

[25] Cellulose. (2008). In *Encyclopædia Britannica.* Retrieved January 11, 2008, from Encyclopædia Britannica Online.

[26] *Chemical Composition of Wood.* ipst.gatech.edu.

[27] Piotrowski, S., Carus, M., May 2011. *Multi-criteria evaluation of lignocellulosic niche crops for use in biorefinery processes.* Nova-Institut GmbH, Hürth, Germany.

[28] Romeo, T., 2008. *Bacterial biofilms.* Berlin: Springer. pp. 258-263. ISBN 978-3-540-75418-3.

[29] Nevell, T. P., Zeronian S. H., 1985. *Cellulose chemistry and its applications.* Ellis Horwood, Chichester York.

[30] Bledzki, A. K., Gassan, J., 1999. Composites reinforced with cellulose based fibres. *Progress in Polymer Science* 24(2): 221-274. DOI: 10.1016/S0079-6700(98)00018-5.

[31] Hon, D. N. S., 1996. *Cellulose and its derivatives. In: Dumitriu S (ed) Polysaccharides in medicinal applications.* Marcel Dekker, New York.

[32] Mohanty, A. K., Misra, M., Hinrichsen, G., 2000. Biofibres, biodegradable polymers and biocomposites: an overview. *Macromolecular Materials and Engineering* 276-277(1), 1-24. DOI: 10.1002/(SICI)1439-2054(20000301)276:1.

[33] Hattallia, S., Benaboura, A., Ham-Pichavant, F., Nourmamode, A., Castellan, A., 2002. Adding value to alfa grass (Stipa tenacissima L.)

soda lignin as phenolic resins. 1. lignin characterization. *Polymer Degradation And Stability* 76(2), 259-264.

[34] Hoareau, W., Trindade, W. G., Siegmund, B., Castellan, A., Frollini, E., 2004. Sugar cane bagasse and curaua lignins oxidized by chlorine dioxide and reacted with furfuryl alcohol: characterization and stability. *Polymer Degradation and Stability* 86(3), 567-576. DOI: 10.1016/j.polymdegradstab.2004.07.005.

[35] John, M. J., Thomas, S., 2008. Biofibres and biocomposites. *Carbohydrate Polymers* 71(3), 343-364.

[36] Ramamoorthy, S. K., Skrifvars, M., Persson, A., 2015. A review of natural fibers used in biocomposites: plant, animal and regenerated cellulose fibers. *Polymer Reviews* 55(1), 107-162. DOI: 10.1080/15583724.2014.971124.

[37] Hon, D. N. S., 1996. *Chemical modification of lignocellulosic material.* Marcel Dekker, New York.

[38] Faruk, O., Bledzki, A. K., Fink, H., Sain, M., 2012. Biocomposites reinforced with natural fibers: 2000-2010. *Prog. Polym. Sci.* 37, 1552-1596.

[39] Cordeiro, N. C., Gouveia, A. G. O., Moraes, A. M., Amico, S. C., 2011. Natural fibers characterization by inverse gas chromatography. *Carbohyd. Polym.* 84, 110-117. DOI: 10.1016/j.carbpol.2010.11.008.

[40] Satyanarayana, K. G., Guimarães, J. L., Wypych, F., 2007. Studies on lignocellulosic fibers of Brazil. Part I: Source, production, morphology, properties and applications. *Compos. Part A* 38, 1694-1709. DOI: 10.1016/j.compositesa.2007.02.006.

[41] Nakatsubo, F., Kamitakahara, H., Hori, M., 1996. Cationic ring-opening polymerization of 3,6-di-Obenzyl-α-D-glucose 1,2,4-orthopivalate and the first chemical synthesis of cellulose. *J Am Chem Soc* 118, 1677-1681. DOI: 10.1021/ja953286u.

[42] Nishimura, M., Takeuchi, Y., De BeUis, L. and Hara-Nishimura, I., 1993. Leaf peroxisomes are directly transformed to glyoxysomes during senescence of pumpkin cotyledons. *Protoplasma* 175, 131-137. DOI: 10.1007/BF01385011.

[43] Kobayashi, S., Kashiwa, K., Kawasaki, T., Shoda, S. J., 1991. Novel method for polysaccharide synthesis using an enzyme: the first in vitro synthesis of cellulose via a nonbiosynthetic path utilizing cellulase as catalyst. *Am. Chem. Soc.* 113, 3079-3084. DOI: 10.1021/ja00008a042.

[44] Balser, K., Hoppe, L., Eicher, T., Wendel M., Astheimer, A. J. Cellulose esters. In: *Ullmann's encyclopedia of industrial chemistry* (Gerhartz, W.; Yamamoto, Y. S., Champbell, F. T., Pfefferkorn, R., Rounsaville, J. F., Eds., Weinheim, New York 1986, pp. 461. VCH. 5th, ed. vol A5.

[45] Crawford, R. L., 1981. *Lignin biodegradation and transformation.* New York: John Wiley and Sons. ISBN 0-471-05743-6.

[46] Updegraff, D. M., 1969. Semimicro determination of cellulose in biological materials." *Analytical Biochemistry* 32 (3): 420-424. DOI:10.1016/S0003-2697(69)80009-6. PMID 5361396.

[47] Rosas, K. A., 2008. *Biocompósitos de almidon termoplastico con microfibras de celulosa.* tesis maestrado, Institutode investigación en ciencia aplicada y tecnología avanzada del instituto politecnico nacional, Altamira. [*Biocompounds of thermoplastic starch with cellulose microfibres.* Master's thesis, Institute of research in applied science and advanced technology of the national polytechnic institute, Altamira.]

[48] Zhang, Y. H. P., Cui, J. B., Lynd, L. R., Kuang, L. R., 2006. A Transition from cellulose swelling to cellulose dissolution by o-phosphoric acid: evidence from enzymatic hydrolysis and supramolecular structure. *Biomacromolecules* 7, 644-648. DOI:10. 1021/bm050799c.

[49] John, M. J., Thomas, S., 2008. Biofibres and biocomposites. *Carbohydrate Polymers* 71, 343-364. DOI: 10.1016/j.carbpol.2007. 05.040.

[50] Moon, R. J., Martini, A., Nairn, J., Simonsen, J., Youngblood, J., 2011. Cellulose nanomaterials review: structure, properties and nanocomposites. *Chemical Society Reviews* 40, 3941-3994. DOI:10. 1039/c0cs00108b.

[51] Labafzadeh, S. R., 2015. *Cellulose-based materials. academic dissertation*, Faculty of Science of the University of Helsinki, Helsinki.

[52] Brown, G., Ley, H. *Science* 1965, 147, 1038.

[53] Chu, S., Jeffrey, G. *Acta Cryst*. 1968, B24, 830.

[54] Berman, H., Kim, S. *Acta Cryst*. 1968, B24, 897.

[55] Ellefsen, O., Tonnesen, B., 1971. *Cellulose and cellulose derivatives*, part IV, John Wiley & Sons, New York, 1.

[56] Rao, V., Sundararajan, P., Ramakrishnan, C., Ramachandran, G., 1957. *Conformation of biopolymers*, Academic Press, London, 721.

[57] https://www.bankofbiology.com/2014/07/comparison-between-starch-glycogen-and.html.

[58] http://slideplayer.com/slide/6217371/.

[59] Battista, O., 1956. *Ind. Eng. Chem.*, 48, 333.

[60] Favier, V., Chanzy, H., Cavaille, J. Y., 1995. Polymer nanocomposites reinforced by cellulose whiskers. *Macromolecules* 28(18), 6365-6367. DOI: 10.1021/ma00122a053.

[61] Favier, V., Canova, G. R., Cavaillé, J. Y., Chanzy, H., Dufresne, A., Gauthier, C., 1995. Nanocomposite materials from latex and cellulose whiskers. *Polymers for Advanced Technologies* 6(5), 351-355. DOI: 10.1002/pat.1995.220060514.

[62] Bishop, Charles A., ed. 2007. *Vacuum deposition onto webs, films, and foils*. p. 165. ISBN 0-8155-1535-9.

[63] Dauenhauer, P., Krumm, C., Pfaendtner, J., 2016. Millisecond pulsed films unify the mechanisms of cellulose fragmentation. *Chemistry of Materials* 28 (1), 0001. DOI:10.1021/acs.chemmater.6b00580.

[64] Wymer, C. E., 1994. Ethanol from lignocellulosic biomass: Technology, economics, and opportunities. *Bioresource Technology* 50 (1), 5. DOI:10.1016/0960-8524(94)90214-3. Retrieved 28 October 2016.

[65] Klock, U., Muniz, G. I. B., Hernandez, J. A., 2005. *Química da madeira. curitiba: fundação de pesquisas florestais do parana-fupef*, Curitiba. [*Chemistry of wood. curitiba: foundation of forest research of parana-fupef*, Curitiba.]

[66] Stenius, P., 2000. "Ch. 1." Forest products chemistry. *Papermaking Science and Technology* Vol. 3. Finland: Fapet OY. p. 35. ISBN 952-5216-03-9.

[67] Wang, H., Gurau, G., Rogers, R. D., 2012. Ionic liquid processing of cellulose. *Chemical Society Reviews* 41 (4), 1519-37. DOI:10.1039/C2CS15311D.PMID 22266483.

[68] Zhou, C.-J., Wu, Q.-L., 2012. Recent development in applications of cellulose nanocrystals for advanced polymer-based nanocomposites by novel fabrication strategies. In book entitled *Nanocrystals - synthesis, characterization and applications,* Neralla, S., (Chap 6), Intech, Croatia, 103-120. DOI: 10.5772/2560.

[69] Xue, L. G., Tabil, L., Panigrahi, S., 2007. Chemical treatments of natural fiber for use in natural fiber-reinforced composites: A review. *Journal of Polymers and the Environment* 15, 25-33. DOI: 10.1007/s10924-006-0042-3.

[70] Mathews, F. L., Rawlings, R. D., 1999. *Composite materials: Engineering ans science woodhead* Publishing Ltd., Cambridge, England.

[71] Eichhorn, S. J., Young, R. J., 2001. The Young's modulus of a microcrystalline cellulose, *Cellulose* 8(3), 197-207. DOI: 10.1023/A:1013181804540.

[72] Saheb, D. N., Jog, J. P., 1991. Natural fiber polymer composites: A review. *Advances in Polymer Technology* 18, 351-363. DOI: 10.1002/(SICI) 1098-2329(199924)18:4.

[73] Mohanty, A. K., Misra, M., Hinrichsen, G., 2000. Biofibres, biodegradable polymers and biocomposites: An overview. *Macromolecular Materials and Engineering* 276-277 (1), 1-24. DOI: 10.1002/(SICI) 1439-2054(20000301)276:1.

[74] Ngo, T. D., Ho, T., 2016. Lignocellulosic biomasses, sustainable platform for bio-based materials with polypropylene. In book entitled *Polypropylene: properties, uses and benefits*, Paula Garcia, Nova Science Publisher. ISBN: 978-1-53610-008-2.

[75] Amandine, C., Sylvain, F., Frédéric, J., Pascal, C., 2014. The hygroscopic behavior of plant fiber: a review. *Frontier In Chemistry* Vol.1, 43, 1-12. DOI: 10.3389/fchem.2013.00043.

[76] Ngo T. D., May 2018. Natural fibres for sustainable bio-composites. In book entitled *Natural and artificial fiber-reinforced composites as renewable sources*, Ezgi Gunay Ed., InTech, Croatia, DOI 10.5772/ intechopen.71012.

[77] Reportlinker, Jan 17, 2017. *Cellulose Fiber Market Analysis By Product Type* (Natural, Synthetic), By Application (Textile, Hygiene, Industrial), By Regions And Segment Forecasts, 2014-2025.

[78] EFS, 2006. *The History of Textiles: Textile fibre usage and production, Education For Sustainability* (EFS), Textile online URL: http://www.e4s.org.uk/textilesonline/content/6library/fr_library.htm [accessed 4 December 2006].

[79] Albrecht, W., 2004. Regenerated cellulose in chapter "Cellulose." In: the book entitled *Ullmann's encyclopaedia of industrial chemistry* 2007, 7th Edition. Wiley-VCH Verlag GmbH & Co. KGaA. DOI: 10.1002/14356007.a05_375.pub2.

[80] Urquhart, A. R., April 1954. *Cellulose acetate as a raw material for rayon production,* 4 (4), 195-203, DOI: 10.1002/jctb.5010040407.

[81] https://en.wikipedia.org/wiki/Pulp_and_paper_industry.

[82] https://en.wikipedia.org/wiki/Pulp_(paper).

[83] http://www.eulerhermes.com/economic-research/blog/Economic Publications/Paper-global-sector-report-feb18.pdf.

[84] https://www.mordorintelligence.com/industry-reports/pulp-and-paper-chemical-market.

[85] https://www.marketsandmarkets.com/PressReleases/biocomposite.asp.

[86] https://www.inkwoodresearch.com/reports/wood-plastic-composites-market/.

[87] Javad, S., Khosro, A., 2013. Application of cellulose and cellulose derivatives in pharmaceutical industries. In book entitled *Cellulose - medical, pharmaceutical and electronic applications,* van de Ven, T.,

Godbout, L., ISBN 978-953-51-1191-7, 314 pages, Publisher: InTech, DOI: 10.5772/55178.

[88] Bethany, M., Updated 01/26/18, *Learn about cellulose and how it is used in food*, https://www.thespruce.com/what-is-cellulose-1328464.

[89] Ardizzone, S., Dioguardi, F., Mussini, T., Mussini, P., Rondini, S., Vercelli, B., Vertova, A., 1999. *Cellulose* 5-20, 57.

[90] Frohoff-Hülsmann, M. A., Lippold, B. C., McGinity, J. W., 1999. Aqueous ethyl cellulose dispersion containing plasticizers of different water solubility and hydroxypropyl methyl- cellulose as coating material for diffusion pellets II: properties of sprayed films. *European Journal of Pharmaceutics and Biopharmaceutics* Volume 48, Issue 1, 67-75.

[91] Frohoff-Hülsmann, M. A., Schmitz, A., Lippold, B. C., 1999. Aqueous ethyl cellulose dispersions containing plasticizers of different water solubility and hydroxypropyl methylcellulose as coating material for diffusion pellets: I. Drug release rates from coated pellets. *International Journal of Pharmaceutics* Volume 177, Issue 1, 69-82.

[92] Rowe, R. C., 1986. The effect of the molecular weight of ethyl cellulose on the drug release properties ofmixed films of ethyl cellulose andhydroxypropylmethylcellulose. *International Journal of Pharmaceutics* Volume 29, Issue 1, 37-41.

[93] Larsson, M., Hjärtstam, J., Berndtsson, J., Stading, M., Larsson, A., 2010. Effect of ethanol on the water permeability of controlled release films composed of ethyl cellulose and hydroxypropyl cellulose. *European Journal of Pharmaceutics and Biopharmaceutics* Volume 76, Issue 3, 428-432.

[94] Delgado, J. N., William A., 1998. *Wilson and Gisvold's textbook of organic medicinal and pharmaceutical chemistry*. Lippincott-Raven Publishers, Wickford.

[95] Fink, H., Faxälv, L., Drotz, K., Risberg, B., Lindahl. T, Sellborn, A., 2009. Real-time measurements of coagulation on bacterial cellulose and conventional vascular graft materials. *Acta Biomaterialia.* DOI: 10.1016/j.actbio.2009.09.019.

[96] https://www.sciencedaily.com/releases/2009/12/091222104907.htm.

[97] https://www.slideshare.net/DRxPaTel/cellulose-derivatives.

[98] Kamel, S., Ali, N., Jahangir, K., Shah S. M., El-Gendy, A. A., 2008. Pharmaceutical significance of cellulose: A review, *eXPRESS Polymer Letters* Vol. 2, No. 11, 58-77. DOI: 10.3144/expresspolymlett.2008.908.

[99] Hercules Incorporated, *Aqualon Division*: http://www.aqualon.com (16.07.2008).

[100] Arthur, J. C. Jr., 1986. In the book entitled *Comprehensive polymer science*, Allen G., Bevington J. C. (Eds.) Vol 6., Pergamon, Oxford.

[101] Marques-Marinho, F. D., Vianna-Soares, C. D., 2013, Cellulose and its derivatives use in the pharmaceutical compounding practice. In book entitled *Cellulose, medical, pharmaceutical and electronic applications,* Theo G. M. Van De Ven, Ed., IntechOpen Limited, London, SE19SG - United Kingdom, DOI: 10.5772/3470, ISBN: 978-953-51-1191-7.

[102] Swatloski, R., Holbrey, J., Spear, S., Rogers, R., 2002. *Electrochem. Soc. Proc.* 19, 155.

[103] Misha M., 2016. *Handbook of encapsulation and controlled release,* CRC Press, Taylor & Francis Group, Boca Raton, FL, USA. ISBN: 9781482232349.

[104] Varshney, V. K., Naithani, S., 2011. Chemical functionalization of cellulose derived from nonconventional sources. In book entitled *Cellulose fibers: bio- and nano-polymer composites*, Kalia, S., Ed., Springer Verlag, Berlin Heidelberg, Germany. DOI 10.1007/978-3-642-17370-7_2.

[105] http://www.bpiexamacademy.com/1-appropriate-insulation-applications-based-on-existing-conditions.html.

[106] https://study.com/academy/lesson/the-biological-function-of-cellulose.html

In: Cellulose to Nanocelluloses
Editor: Tri-Dung Ngo
ISBN: 978-1-53615-057-5
© 2019 Nova Science Publishers, Inc.

Chapter 2

NANOCELLULOSE: FROM PRODUCTION TO APPLICATIONS

***Tri-Dung Ngo*[1,2,*], *Christophe Danumah*[1] and *Behzad Ahvazi*[1,3]**
[1]InnoTech Alberta Inc., Edmonton, Alberta, Canada
[2]University of Alberta, Faculty of Engineering, Department of Civil & Environmental Engineering, Edmonton, Alberta, Canada
[3]University of Alberta, Faculty of Engineering, Department of Mechanical Engineering, Edmonton, Alberta, Canada

ABSTRACT

Preserving the environment and continuous demands for "bio" and "green" materials has been growing considerably for the past several decades. The use of sustainable and bio-degradable resources have attracted the attention of many scientists to investigate and develop applications of bio-based products.

* Corresponding Author Email: tridung.ngo@innotechalberta.ca.

This chapter highlights the roadmap from cellulose to its own different derivatives at nanoscale using agriculture- and forest-based products. This paves the way not only to diversify the renewable agriculture and forest product industry through nanotechnology but also to engineer and develop applications of novel nano-based biomaterials.

Keywords: cellulose, cellulose nanocrystals, cellulose nanofibrils, productions, applications

INTRODUCTION

In recent years, the conversion of renewable lignocellulosic biomass and natural biopolymers into chemicals, liquid fuels and feed supplements has gained considerable attention [1]. This is mainly due to the high-energy intensity in production of chemicals and synthetic materials. With appropriate conversion and extraction technologies as well as modification and characterization, biopolymers like cellulose can be integrated into bio-based products. Their use in novel materials and various applications favours the future usage of lignocellulosic-based biomass components with substantial environmental and economic benefits [1, 2].

Research centers focusing on biomass conversion and processing play a strategic role in bridging forest industries by transforming wood and crop waste into chemicals with commercial applications. The goal is to reduce the world's reliance on oil, decrease our environmental footprint, and also grow the local economy. Dedicated research scientists at these centers are experimenting with residue from crops and wood products to find new environmentally sustainable products made with cellulose. This could help revitalize agriculture and forestry industries and pave the way toward providing novel bio-based greener materials for use in the energy sector and many other industries.

The ultilization of the cellulose based materials continues today as demonstrated by the enormity of the global industries such as forest products, paper, textiles, composites, etc. However, their consistency, uniformity, properties, functionality, and durability are still limited.

Therefore, the next generation of cellulose-based products will be required for novel engineering applications which cannot be achieved with traditional cellulosic materials [3].

Nanotechnology is a rapidly evolving technology and as science, engineering and technology have merged to bring nanoscale materials that much closer to reality. This is one of the fastest emerging areas for research [4, 5]. Nanofibers, nanotubes, nanoparticles, nanowhiskers, nanoplates and so on, are just substances that are within the nanoscale (less than 100 nanometers) in at least one dimension. Too small to see but too big to ignore, nanotechnology is helping improve products that we use every day and creating new, exciting products for the future. Since 2000, there has been increasing interest from different corporations, governments, research organizations, universities and the public sector to demonstrate that nanotechnology has the potential to improve and diversify forest-based products for new cellulose-based applications with different properties. To completely change the forest products industry through nanotechnology, top-down approaches such as chemical, physical, biological, mechanical or a combination of each of them have been used to generate cellulose at nanoscale, so called nanocellulose, such as cellulose nanofibrils (CNF) and cellulose nanocrystals (CNC). With appropriate conversion and extraction technologies as well as modification and characterization, cellulose can be converted to nanocellulose like CNF and CNC [6, 7]. Both CNF and CNC are very attractive because of their intriguing properties such as their high strength and stiffness combined with low weight, biodegradability and renewability, low thermal expansion, low toxicity, and anticipation of low cost. Their utilization in novel materials and various applications favours the future applications of cellulose and cellulose-based components with substantial environmental and economic benefits [8].

NANOCELLULOSE

Nanocellulose could be extracted from almost any cellulosic materials. Cellulose is the most important and naturally abundant organic biopolymer in the biosphere. The nanocelluloses have been isolated from a variety of cellulosic sources, including plants, wastes, animals (tunicates), bacteria and algae. Plant sources cover the wood of forest resources, and nonwood lignocellulose. Wood is a porous and fibrous structural tissue which is found in the stems, trunk, branches, roots of trees and other woody materials. Wood is an organic material, a natural composite of hard cellulose fibers that are strong in tension and embedded in a matrix of lignin that resists compression. Wood plants including resources as softwood and hardwood have many advantages as cellulose sources. Most significantly, it is abundant and is renewable. In addition, an existing infrastructure is in place for harvesting, processing and handling, i.e., the pulp and paper, packaging and pharmaceutical industries. Plants such as coir, cotton, hemp, flax, jute, kenaf, ramie, sisal, etc. are attractive cellulose sources primarily because they are abundant and there is a preexisting infrastructure in the textile industries for harvesting, retting/pulping (i.e., to treat and isolate micron sized cellulose particles), and product processing. In general, plants can be "purified" similarly to wood.

Apart from the plant-based source, nanocellulose can also be obtained from microorganisims, especially bacteria such as Acetobactor, Agrobacterium, Rhizobium, etc. The most obvious choice of nanocellulose sources is wood. Wood, owing to its natural abundance, is a key source of cellulose because of its widespread availability and high content of cellulose [9,10].

Moving from cellulose to cellulose fibers at the nanoscale will allow these renewable materials to be utilized to get high value-added products with special performance and reach new markets. For instance, cellulose is present in the cell wall of wood fibers along with hemicellulose and lignin. The cellulose fibers are composed of microfibrils that, in turn, are composed of elementary fibrils, which are the basic structural units.

In the cellulose fibers, the elementary fibrils or nanofibers are about 2-20 nm in diameter and a few micrometers in length. About 30 to 100 cellulose chains aggregate into an elementary fibril. There are regions within each of these elementary fibrils, where the cellulose chains are arranged in highly ordered structures (crystalline regions) and regions that are disordered structures (amorphous regions). These elementary fibrils are formed during cellulose biosynthesis. Figure 1 shows the schematic illustration of cellulose fiber to nanocellulose.

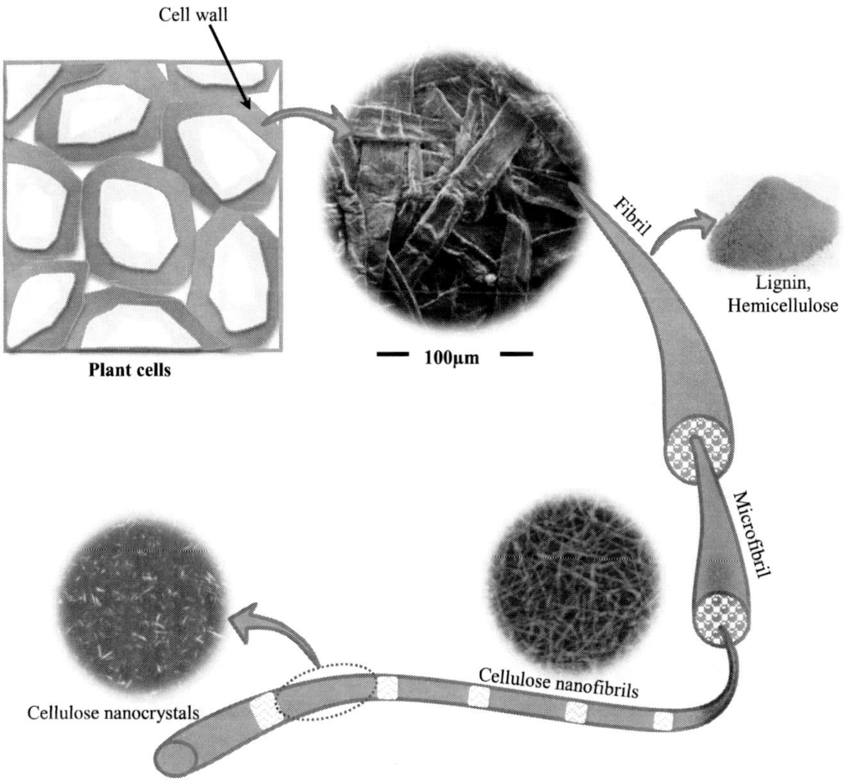

Figure 1. Cellulose fiber to nanocellulose.

Cellulose is the most ubiquitous and naturally abundant organic biopolymer in the biosphere. It is comprised of carbon, hydrogen, and oxygen bioengineered as a linear homopolysaccharide chain with cellobiose as the repeating building blocks [1,11,12]. Each block is composed of two adjacent anhydro-glucose rings connected via a β-1,4 glycosidic linkages as illustrated in Figure 2.

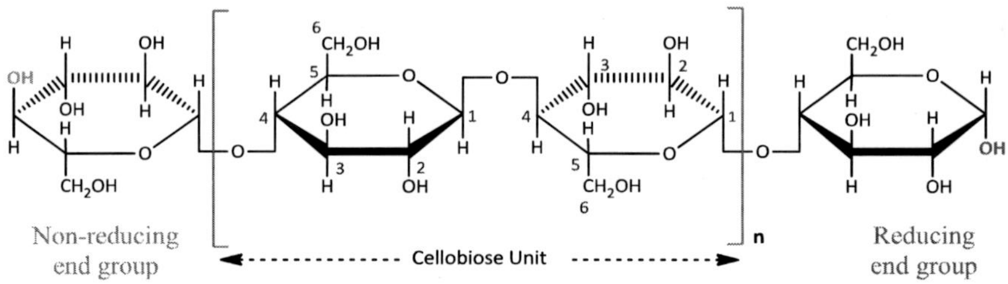

Figure 2. Schematic diagram of partial molecular structure of cellulose chain with numbering for carbon atoms and n=number of cellobiose repeating unit.

The cellulose chain contains two different OH groups at its both ends. Though there are OH groups at both ends of the cellulose chain, these OH groups show different behavior. The C_1-OH is an aldehyde group resulting from the ring formation by an intermolecular hemiacetal linkage. That is why the OH group at the C_1 end has reducing properties, while the OH group at the C_4 end of the cellulose chain is an alcoholic hydroxyl with no-reducing properties.

The degree of polymerization (DP) of glucose units in cellulose varies from 7,000-10,000. The presence of a large number of hydrogen bonds results in a strong lateral association of linear cellulose molecules [13]. This strong association and almost perfect alignment of the cellulose molecules gives rise to crystallinity for every 60 nm followed with non-crystalline or amorphous regions. In the amorphous region, the molecules are arranged in an irregular and non-periodic manner. Therefore, the cellulose molecules can be considered as highly oriented (crystalline), then pass through an area of poor orientation (amorphous) and re-enter a crystalline region. The pattern repeats throughout the length of the cellulose molecule.

Cellulose is odourless, flavourless, hydrophilic and is insoluble in most solvents, including strong alkali. It is biodegradable and can be broken down chemically, in addition to other techniques, into its monomeric building blocks for production of biofuels.

Nanotechnology draws from different fields including biology, chemistry, chemical engineering, colloid science, electrical, forestry, material science, and physics. Nanocellulose is a general nomenclature related to cellulose particles with at least one dimension in the nanoscale and can be assumed independent of the sources, the production methods as well as the morphologies [14]. The research on nanosized bio-based cellulosic materials, nanocellulose, has been scrutinized in recent years because of the unique combination of their properties including outstanding mechanical properties, surface chemistry, nontoxicity, biocompatibility, and most importantly, their abundance from renewable and sustainable resources.

There is a confusing litany of terms that many cellulose materials are referred to as nanocellulose such as nanocrystalline cellulose (NCC), cellulose nanocrystals (CNC), cellulose whiskers, nanofibrillated cellulose (NFC), cellulose nanofibrils (CNF), microfibrillated cellulose (MFC), carboxymethylated cellulose (CMC), microcrystalline cellulose (MCC), and cellulose filaments (CF). However, by definition, nanocellulose is nano-scale materials which have at least one dimension less than 100 nanometers in size. Two materials fitting this definition are cellulose nanocrystals (CNC), sometimes called nanocrystalline cellulose (NCC) or cellulose nanowhiskers, and cellulose nanofibrils (CNF), sometimes called nanofibrillated cellulose (NFC).

PRODUCTION

The production of nanocellulose such as CNC and CNF is an emerging technology that has gained considerable interest from a range of industries that are working towards greater utilization of bio-based materials. The proportions of cellulose vary depending on the material from which it is sourced. Figure 3 presents the pathways to nanocelluose from cellulose-based materials.

Breaking down the cellulose to the elementary fibrils and further to crystalline regions and generate the nanocellulose such as CNF and CNC will provide the value-added products to the cellulose [15,16]. Cellulose nanofibrils (CNF) are materials containing fibrils with length in the micrometer and width in the nanometric range, forming a network structure. The nanofibrils contain both amorphous and crystalline cellulosic regions [17]. CNF can be prepared by its liberation from the constituent fiber and microfiber bundles [18, 19, 20]. Various approaches have been utilized to prepare CNF materials with different degrees of deconstruction. The processes lead to materials with sizes in the ranges of micro to nano-scales. CNF has been produced from wood since the 1980s. When this material was first introduced, it was referred to as microfibrillated cellulose (MFC).

CNF can be produced by mechanical procedure with or without pre-treatment from any cellulosic-based materials including wood pulp, recycled paper and paperboard, cotton fibers, hemp, flax, bamboo, sugarcane bagasse and other agro-biomass. Depending on the mechanical force levels and types of mechanical treatment, interfibrillar hydrogen bonding are broken.

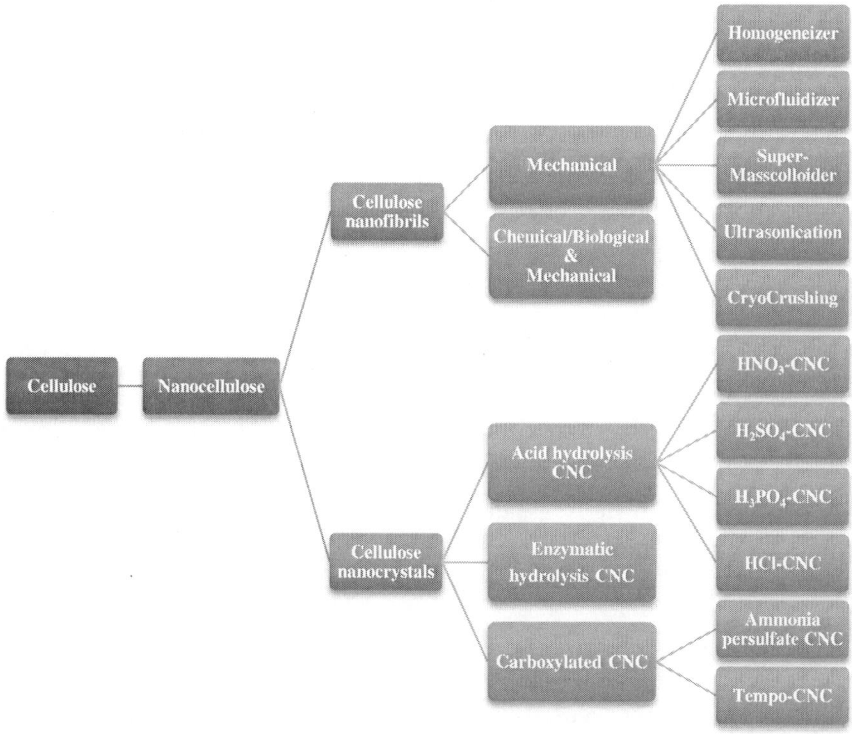

Figure 3. Pathways from cellulose to nanocelluloses.

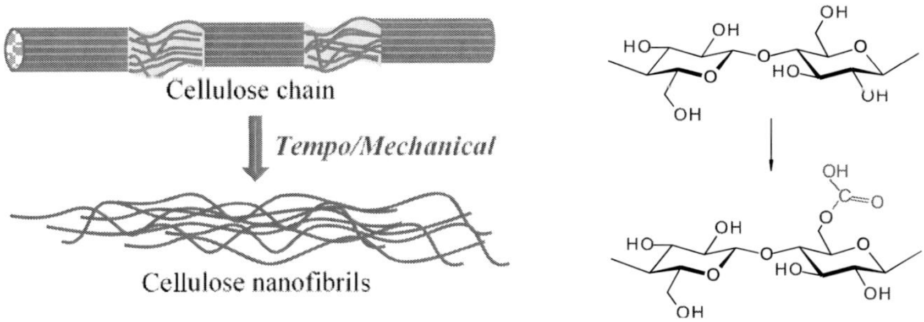

Figure 4. Formation of CNF and carboxylate groups onto CNF surface through Tempo/mechanical process.

Microfluidization and high-intensity ultrasonic treatments produce a high shear gradient causing transverse cleavage along the longitudinal axis of the cellulose fibers, and as a result, they tend to damage the microfibril structure by reducing the molar mass and degree of crystallinity. The mechanical methods exhibit high production costs (tools and materials) since the fibrillation of pulps by mechanical means is a highly energy demanding process (between 30,000kWh/ton and 70,000kWh/ton). They then require greater energy and they are less efficient than the chemical methods [21]. This explains why high yield or ultra high yield pulp is often used as feedstock to reduce operational cost. During this process, biological or chemical pre-treatments can be used as preprocessing steps to weaken the hydrogen bonding and facilitate the mechanical processing and decrease the energy consumption to 1000kWh/ton. The most commonly used approaches to isolate CNF are enzymatic treatment and Tempo oxidation followed by mechanical defibrillation. The Tempo converts the primary hydroxyl groups (C6) to charged aldehyde or carboxylate functional groups, whereas the secondary hydroxyl moieties present in the cellulose molecule remain unaffected [22, 23]. Figure 4 presents the formation of CNF and carboxylate groups onto CNF surface through Tempo/mechanical process.

Figure 5 presents images of a microfluidizer and a supermasscolloider which can be utilized for CNF production. The microfluidizer family of high pressure homogenizers excels at efficiently converting extreme fluid pressures into high shear forces more efficiently and consistently than is readily achievable using competing technologies. Results are remarkably consistent because microfluidizer processors ensure that by maintaining constant pressure, a given material receives exactly the same treatment [24]. Supermasscolloider, which is even called "Dissolving machine", can make ultra-fine particle products which look as if they are dissolving and the machine features two ceramic non-porous grinding stones. Clearance between these two stones can be adjusted freely. While the upper grinding stone if fixed, the lower one is rotated at a high speed. Raw materials fed into the hopper are dispersed by centrifugal force into the clearance between those grinding stones, where they are ground into ultra-fine particles, after being subjected to massive compression, shearing, and rolling friction forces. As is a stone-mill type machine, it produces more round shape and smoother particles than other types of crushing machines and is the most popular of small production models, as it takes up as little installation space as one square meter. All models feature jackets to accommodate cooling and heating applications. They treat a wide range of raw inputs from low to high viscosity materials [25].

Figure 5. Images of microfluidizer (left) [26] and supermasscolloider (right) [27].

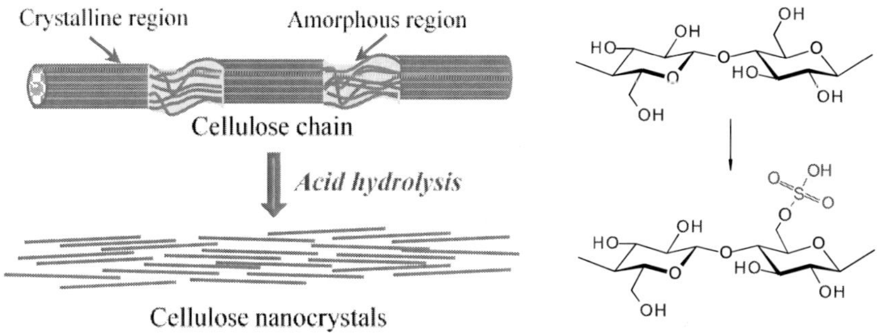

Figure 6. Formation of CNC and sulfate groups on CNC surface.

CNC can be prepared by chemical treatments such as acid hydrolysis, oxidation, or biological treatment as enzymes cleave the amorphous regions of the cellulose fibers and liberate CNC. The first report regarding the production of cellulose nanocrystals in solution is dated from the beginning of the 1960s and corresponds to the paper of Rånby and Noe [28]. Acid hydrolysis process is the predominant method used to isolate CNC and sulfuric acid hydrolysis is the most well-known and efficient. By treating cellulose with concentrated acids, the amorphous regions are attacked, leaving the crystalline regions intact, thereby producing nano-sized cellulose-based crystals called NCC or CNC.

A typical CNC production process involves acid hydrolysis, quenching, centrifugation, dialysis, and homogenization to generate purified CNC suspensions followed by drying [1, 29]. The main process for the preparation of CNCs is based on strong acid hydrolysis under controlled conditions of acid concentration, acid-to-cellulose ratio, reaction temperature, agitation, and time. The acid breaks the β-1,4 glycosidic bonds in cellulose. The resulting suspension is neutralized, centrifuged, and dialysis removes any impurity molecules. When sulfuric acid is used, anionic groups (half-ester sulfonates) are formed on the CNC nanorods surface as shown in Figure 6, helping their stability in aqueous suspension.

Significant research has focused on understanding and optimizing CNC isolation and it has been found that changing the hydrolysis conditions can greatly impact particle dimensions, surface charge density and overall CNC yield. Figure 7 presents an example of major pilot plant components such as

1) feedstock, 2) reactors, 3) centrifuge, 4) microfiltration modules, 5) spray dryer and 6) packaging for CNC production. Feedstock and acid hydrolysis reaction conditions are the two key components that govern the overall operation to achieve high yields and consistent production of quality CNC that is cost effective.

It is critical that the community as a whole recognizes that CNC produced from various sources and methods may not have the same properties and behaviour. The various hydrolysis conditions, post extraction treatments, and diversity of cellulose starting materials highlight the need for thorough characterization of CNC prior to use. The Canadian Standards Association (CSA) has outlined standard test methods to assess cellulosic nanomaterial properties, and current Standards and nomenclature guidelines are in preparation by ISO and TAPPI, however, there is yet to be direct comparison between industrial and lab scales CNC [30].

Figure 7. Schematic diagram of major CNC pilot plant units.

Table 1. Global CNF and CNC production capacity [31, 32, 33]

	Organization	Country	Production Capacity
CNF	American Process Inc.	USA	500 kg/day, 100 tons/year
	AkzoNobel	Netherlands	Pilot scale
	BASF	Germany	Pilot scale
	Borregaard	Norway	Pilot scale 500/day, 150 tons/year. Commercial scale 1000 tons/year
	Brazilian Agricultural Research Corporation (Embrapa-Forestry)	Brazil	Pilot plant
	CelluComp	UK	Biorefinery plant 15 to 400 tons/day. New plant planned 50 tons/year
	Cellutech AB	Sweden	-
	Daicel	Japan	Lab
	Daio Paper Corp.	Japan	Pilot plant 100 tons/year, 1000 tons/year planned
	DKS Co., Ltd.	Japan	50 tons/day
	Domtar	Canada	Planned
	FPInnovations	Canada	Pilot
	FiberLean Technologies	UK	8000 tons/year
	Fibria	Brazil	2 tons/day
	India Council of Agricultural Research Central Institue for Research on Cotton Technology (ICAR-CIRCOT)	India	Pilot plant 10kg/shift up to 210 kg/week
	Innventia	Sweden	100 kg/day
	InoFinb	France	Lab scale
	InTechFibres	France	Pilot sacle 100 kg/day
	Klabin	Brazil	Production planned
	Kruger Biomaterials Inc.	Canada	6000 tons/year
	Luleä University of Technology	Sweden	Lab
	Namicell	France	100 kg/day
	Nippon Paper	Japan	150 kg/day
	Norske Skog	Norway	Pilot plant 1 ton/day
	Oji Paper	Japan	100 kg/day
	Performance BioFilaments Inc.	Canada	Commercial production planned for 2018 (50k to 100k tons/year)
	Renmatix	USA	10 kg batch size
	Rise Bioeconomy	Sweden	Pilot plant 100 kg/day. Demo plant 1 ton/day
	Sappi	Netherlands	Pilot scale 8 tons/year
	Stora Enso	Sweden and Finland	Pilot

Table 1. (Continued)

	Organization	Country	Production Capacity
CNF	Suzano	Brazil	50 kg/day
	Turners Falls Paper	USA	2 tons/day
	UPM	Finland	Pilot plant and demo plant
	University of Alberta/InnoTech Alberta	Canada	Pilot scale 5kg/day
	Universidade Federal de Viçosa	Brazil	Pilot plant
	University of Maine	USA	1 ton/day
	University of Queensland	Australia	Pilot plant 5 kg/day
	USDA Forest Service, Forest Products Lab	USA	100 gallon reactors
	VTT collaboration with Aalto U, UPM	Finland	Semi-pilot plant
	Weidmann	Switzerland	500 tons/year
CNC	Advanced Cellulosic Material Inc. (formerly Biovision Technology Enterprises Inc.)	Canada	20 kg/week
	American process Inc.	USA	500 kg/day
	Blue Goose Biorefineries Inc.	Canada	10 kg/day, 35 kg/week
	Brazilian Agricultural Research Corporation (Embrapa)/National Nanotechnology Laboratory for Agriculture	Brazil	Piolot plant
	Caspian Nanocellulose Polymer Development Co.	Iran	-
	CelluForce	Canada	1 ton/day
	Colorado School of Mines	USA	Lab
	Indian Council of Agricultural Research - Central Institute for Research on Cotton Technology (ICAR-CIRCOT)	India	Pilot plant 10 kg/shift up to 210 kg/week
	InnoTech Alberta/Alberta-Pacific Forest Industries Inc.	Canada	up to 20 kg/day
	Innotech Materials LLC	USA	20 kg/week
	Guilin Qihong Technology	China	-
	FPInnovations	Canada	3 kg/day
	Melodea Ltd.	Sweden	100 kg/day
	Melodea Ltd.	Israel	Pilot plant
	Tianjin Haojia Cellulose Co./Tianjin Woodelf Biotechnology Co., Ltd.	China	Pilot scale
	US Forest Service, Forest Products Lab (FPL)	USA	10 kg/day; ca. 90 kg/week

There is a worldwide research effort to reduce the cost of nanocellulose by increasing the production to an industrial scale and there are already initiatives in Finland, United States and Canada to produce CNC at pilot scale. Nanocellulose has enormous potential to provide an important platform to many sectors. While Europe and Asia are leading in the development and commercialization of CNF, North America (US and Canada) is more focused on CNC production and applications development, even if CNF manufacturing seems to be more cost-effective than CNC. FPInnovations estimates the market for nanocellulose to be worth $250 million in North America by 2020. The U.S. National Science and the National Nanotechnology Initiative estimated the global potential market for products using nanotechnology at nearly US $3 trillion by 2020, with wood products estimated to capture 20% of this market. Table 1 provides global production capacity for CNF and CNC.

PROPERTIES

Nanocellulose is a unique and promising natural material extracted from cellulose. It has remarkable physical properties, special surface chemistry and excellent biological properties (biocompatibility, biodegradability and low toxicity). The nanocellulose materials are being regarded as the next generation of renewable reinforcements for the production of high performance biocomposites. CNF and CNC can be isolated from same cellulose-based sources. Both CNF and CNC are two types of nanocellulose with very similar chemical compositions, but significant morphological differences with specific properties depicted in Table 2. Their dimensions vary depending on the native cellulose sources, extraction methods and recovery processes (e.g., hydrolysis time and temperature).

The similarities and differences make CNF and CNC suitable for producing high-quality, durable, light weight and cost effective products for a variety of applications. Besides the "green" advantage of CNF and CNC,

Table 2. Similarities and differences between of CNF and CNC

	CNF	CNC
Similarities	• Nanoscale cellulose • Renewable, recyclable, biodegradable and biocompatible • High surface area • Excellent mechanical properties • Low density • Low thermal expansion	
Differences	• Isolated by mechanical, ultrasonication, and chemi/ or bio/mechanical • Flexible with both amorphous and crystalline regions, roughly 5-100nm in diameters and micrometer or more in lengths • High degree of polymerization • Medium crystallinity (<70%) • High water binding capacity than CNC • Not stable suspension in water at low consistency, and gel at 1-2wt% consistency • No self-assembly • High aspect ratio (>100)	• Isolated by acid hydrolysis, enzymatic, carboxylated, oxidation • Crystals, needle- (or rod-) shaped particles with 5-70nm in diameters and 100 to 300nm in lengths • Low degree of polymerization • High crystallinity (>80%) • Lower water binding capacity than CNF • Form stable suspension in water at 1-3wt% consistency and gel at 8-10wt% consistency • Self-assembly • Aspect ratio (30-100), lower than CNF

they have the potential to be processed on an industrial scale at lower cost than other nanomaterials. CNF can be considered "noodle-like" and CNC is more "rice-like" as shown in Figure 8. The figure also shows SEM images in transmission mode of the CNF and CNC. Each CNC or CNF has very specific properties (e.g., mechanical, high water binding capacity, self-assembly and optical) which are expected to complement one another in certain areas of applications. Some applications rely on the use of CNF produced by mechanical processes, whereas other applications employ CNC. Due to the unique differences in morphology, CNC will form a gel at high consistency (~8-10wt %) as compared to CNF at low consistency (~1-2wt %).

CNF is a material composed of nanosized cellulose fibrils of crystalline and amorphous regions. CNF has a high aspect ratio (length-to-width ratio) and is more flexible than CNC. Typically, the diameters and lengths are 5-

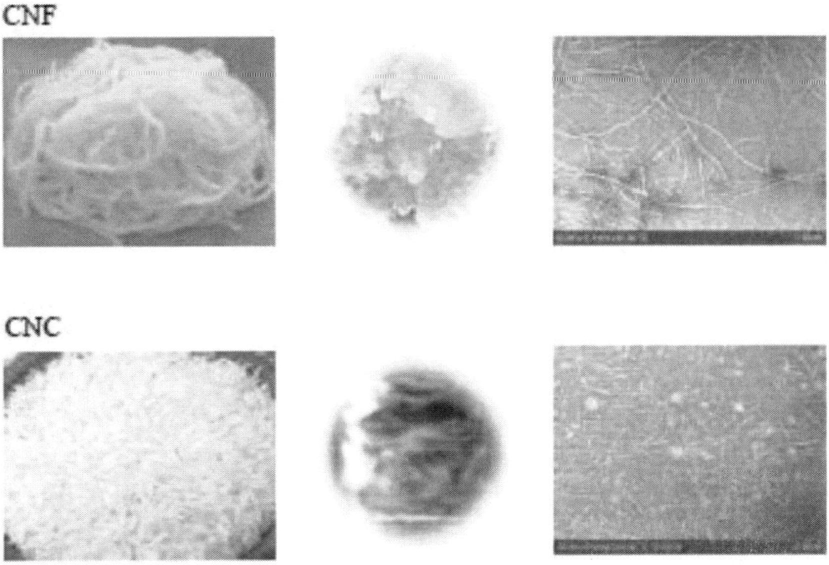

Figure 8. CNF versus CNC apprearances [34,35,36].

100nm and several micrometers, respectively. By hydrolysing cellulose with concentrated acids, the amorphous regions can be broken up, thereby producing cellulose nanocrystals. CNC are rod-like pure cellulose crystals with 5-70nm in diameters and 100-300nm in lengths. Lengths and widths as well as properties of CNC materials highly depend on hydrolysis conditions (acid types, reaction temperature, acid concentration, etc.) and the source of cellulose.

Table 3. Properties of nanocellulose as compared to commonly used synthetic fibers

Fiber	Density (g/cm³)	Elongation (%)	Tensile strength (MPa)	Young's modulus (GPa)
CNC	1.6		7500	120.0-220.0
CNF	1.5-1.6		2000	150.0
Kevlar	1.4	3.3-3.7	3000-3150	63.0-67.0
E-glass	2.5	2.5-3.0	2000-3500	70.0
S-glass	2.5	2.8	4570	86.0

The rheologial behavior of the nanocellulose materials can be complex but shows a great ability to tailor shear thinning behavior. The decrease in the viscosity with increasing shear rate is useful in applications like paints in which a certain flowability is required to allow the paints to better wet and coat the target surface; then once applied, the paints regain their high viscosity and flow stops, and therefore no dripping. Rheologial property is fundamental for the application of nanocellulose as they are produced as aqueous suspensions, and incorporated into various solvents and polymers in the manufacturing of composites. They can be utilized as rheology modifiers such as in adhesives, cosmetics, drilling fluids, and paints [37, 38, 39,40].

Although the production of nanocellulose is first dated from the beginning of 1960s, it is over the 1990s that the remarkable properties of nano-sized cellulose entered researchers' keen again, especially as filler in nanocomposites for mechanical properties improvements of various natural and synthetic polymers [41, 42]. Table 3 shows information about mechanical properties data of CNC and CNF as compared to properties of commonly used fibers such as glass fiber, and aramid fiber [43, 44, 45, 46]. In addition, CNC and CNF offer high tensile strength and can be compared to Kevlar fiber in stiffness, making them excellent materials for reinforcement of nanocomposites [47, 48].

One unique optical property for nanocellulose comes from the liquid crystalline nature of nanocellulose suspensions, which allows nanocellulose to self-assemble into nematic and chiral nematic structures, especially with CNC suspensions at specific concentrations. These highly ordered structures can also persist in a final composite, resulting in interesting optical phenomena. The nanoscale diameters of nanocellulose result in lower light scattering as compared to the micron-to millimeter size of pulp fibers, making good transmittance and low-haze nanocellulose-composites possible. When the nematic structure of the nanocellulose is preserved, the highly aligned CNC materials result in transparent composites and when the chiral nematic structures are obtained, an irridescent/pearlescent optical behavior is observed. Therefore, it is possible to control the color of the dried composite films as well as optical properties [29, 49, 50, 51].

Another property of nanocellulose is its ability to form interconnected network structures in suspension through a self-assembly process. This network arrangement is dependent on the type, morphology, surface chemistry, suspension characteristics of the nanocellulose, input energy, and solvent properties. These network structures can have much different properties as compared to what can be produced with molecular cellulose or pulp. The network structures can be grouped into four categories such as (I) interconnected fiber network structures, (II) dense aligned fiber structures, (III) chiral nematic structures, and (IV) aerogels [29, 52]. The properties can open possibilities of utilizing the nancellulose in different applications such as antimicrobial applications, areas of biosensors, barrier films, composite, coating, catalysis, cement, drug delivery, filters, photovoltaics, and others.

APPLICATIONS

The applications development of nanocellulose has experienced significant growth in recent years because of the unique combination of their properties including outstanding mechanical properties, rich surface chemistry, nontoxicity, biocompatibility, anisotropic shape, tailorable surface chemistry, interesting optical properties and availability from renewable sources. In addition, increased concern about the environment and demands for sustainability in products is driven by consumers and the market push for greater sustainability in the use of products and services. This is leading to the development and use of products from renewable resources and is becoming one of the central priorities of the coming decades.

Nanocellulose has enormous potential to provide an important platform to many sectors. A wide range of industries such as automakers, major branch companies, paper and paper packaging, textiles, cement, and pharmaceuticals companies are highly interested in CNF and CNC for their applications in sectors such as composites, filtration, medical and life sciences, aerospace, military consumer goods, electronics, sensors, concrete, adhesives, paints, etc. [53]. Tables 4 and 5 show the potential applications of CNF and CNC with their associated volumes. The potentially largest volume of cellulose nanomaterials include those applications involving paper and paper packaging, textiles, cement, and automobile parts. Smaller volume applications include sensors, construction, aerospace materials, cosmetics, pharmaceuticals, and paint additives. Novel applications are innovations without current markets and may employ the electrical and photonic properties of cellulose nanomaterials. Additive manufacturing (3D printing) may become a very large volume user of cellulose nanomaterials for toys, architectural models, and parts, but more research is required before it can be ready for commercialization.

Table 4. Application demands on different areas for nanocellulose [54]

High Volume Applications	Low Volume Applications	Novel and Emerging Applications
Automotive body	Aerogels for the oil and gas industry	3D printing
Automotive interior	Aerospace interiors	Air filtration
Cement	Aerospace structure	Cosmetics
Hygiene and absorbent products	Insulation	Excipients
Packaging coatings	Paint-architectural	Flexible electronics
Packaging filler	Paint-OEM applications	Organic LED
Paper coatings	Paint-special purpose	Photovoltaics
Paper filler	Wallboard facing	Photonic films
Plastic film replacement		Purification
Replacement-plastic packaging		Recyclable electronics
Textiles for clothing		Reinforcement fiber-construction
		Sensors-medical, environmental, industrial
		Viscosity modifiers
		Water filtration

The potential applications for nanocellulose has been reported by Cowie and his team as shown in Table 5 [55]. This market study was performed based on the gross domestic product (GDP) and with the assumption that nanocelluloses are at near-term potential for commercialization. Total market sizes for the new products containing nanocellulose materials were estimated from available data pertaining to the existing targeted products, with estimates provided for both high and low volume applications. The report also pointed out that several low volume and novel application categories either have no analogous products existing today or have no market data available to support a market size estimate. Therefore, these products exhibit the highest uncertainty in estimates of market size and degree of market penetration, and are not included in the estimates.

Table 5. Estimated volumes for utilization of nanocellulose for different industries [47]

Applications	Pessimistic estimate		Reasonable estimate		Optimistric estimate	
	%	metric tons (thousands)	%	metric tons (thousands)	%	metric tons (thousands)
High volume						
Automotive body	15	2,700	20	3,573	50	8,950
Automotive interior	15	420	20	587	25	700
Cement	-	-	25	4,130	50	8,250
Hygiene and absorbent	15	1,944	25	3,241	50	6,481
Packaging coatings	15	3,958	20	5,278	25	6,297
Packaging filler	15	1,437	25	2,394	35	3,352
Paper coatings	15	1,625	20	2,167	35	2,708
Paper filler	15	1,437	25	2,394	35	3,352
Plastic film	5	1,683	10	3,366	15	5,049
Plastic packaging	1	1,806	2.3	4,153	4	7,222
Textiles	25	1,273	50	2,543	75	3,891
TOTAL	-	18,283	-	32,829	-	56,481
Low volume						
Aerogels for the oil and gas industry	25	694	50	1,389	75	2,083
Aerospace interiors	50	23	75	32	100	46
Aerospace structure	10	3	30	8	50	13
Architectural paint	10	23	25	56	40	93
Insulation	2	222	5	556	7	778
Paint for OEM applications	50	28	75	56	100	81
Special purpose paint	50	56	75	83	75	111
Wallboard facing	25	25	50	51	75	76
TOTAL	-	1,075	-	2,230	-	3,352

Table 6. Market assessment for utilization of CNF and CNC (x10^3 tonnes) [56]

Application Areas	Market size	Nano-cellulose potential	Potential @ 5% market penetration	CNF (%)	CNC (%)
Adhesives	4,000	80	4	-	95
Cements	15,000	75	4	5	95
Composites	9,000	180	9	5	95
Excipients	4,600	92	5	10	90
Films and barriers	9,670	193	10	-	100
Manufactured textiles	56,300	1,126	56	-	100
Natural textiles	34,500	690	35	-	100
Nonwovens	7,000	140	7	5	100
Oil and gas	17,500	175	9	10	90
Paints and coatings	40,000	800	40	5	95
Paper and paperboard	400,000	20,000	1,000	95	5
TOTAL		23,551	1,178		

However, there are very few cellulose nanomaterial-enabled products that currently exist in the marketplace. According to Cowie, the potential market size for cellulose nanomaterials was determined from complementary goods that are targeted for displacement by cellulose nanomaterials enhanced products. Moreover, the nanocellulose (CNC and CNF) materials are expected to complement each other in certain areas of applications and will impact the forest industry because of its distinctive capacity to be integrated into other materials, as well as its high strength, high surface area and aspect ratio. Therefore, other estimates of the potential application based on the CNC and CNF was reported by Miller as shown in Table 6.

REMARKS

Innovations toward more bio-based, environmentally friendly products are a growing trend, influencing business decisions and corporate investment globally.

The attractive properties of nanocellulosic materials such as biodegradability, biocompactibility, renewability, low density, high strength, good stiffness, low thermal expansion, and high aspect ratio make them suitable for various applications. The nanocelluloses will be used in a wide range of applications and allow the creation of high value products.

The future for nanocellulose is bright, but it will require hard and careful work to demonstrate the many benefits from that most enigmatic of natural polymers, cellulose and its derived nanocellulose. A great effort has still to be made to reduce the high cost associated with the production process of nanocellulose. In addition, thinking outside the box and exploring new and unusual applications may provide for a quicker adoption of the nanocellulose (CNC and CNF)-based technology in a world that is rapidly becoming greener and more sustainable and turning to cellulose and other natural products for answers.

REFERENCES

[1] Ngo, T.-D., Danumah, C., Ahvazi, B., 2018. Production of cellulose nanocrystals at InnoTech Alberta. In the book entitled *Nanocellulose and sustainability: Production, properties, applications, and case studies*, Lee/ Koon-Yang, CRC Press, Taylor & Francis, Boca Raton, FL, USA, ISBN 9781498761031.

[2] Chandel, A.K., Silvério da Silva, S., 2013. *Sustainable degradation of lignocellulosic biomass - techniques, applications and commercialization.* InTech, Janeza Trdine 9, 51000 Rijeka, Croatia. DOI : 10.5772/1490.

[3] Moon, R.J., Marttini, A., Nairn, J., Simonsen, J., Youngblood, J., 2011. Cellulose nanomaterials review: structure, properties and nanocomposites. *Chem. Soc. Rev.*, 40, 3941-3994. DOI: 10.1039/c0cs 00108b.

[4] Elcosh.org/document/1937/d001104/nanotechnology-and-construction.html.

[5] www.fundacionbankinter.org/documents/20183/137602/Publicacion +PDF+IN+FTF_Nanotecnologia.pdf/ca7d4f6e-e2c9-49d4-b6ee-c2200d2926b0.

[6] Mishra, R.K, Sabu, A., Tiwari, S.K., 2018. Materials chemistry and the futurist eco-friendly applications of nanocellulose: Status and prospect. *Journal of Saudi Chemical Society*, DOI: 10.1016/j.jscs.2018.02.005.

[7] Miller, J., 2015. *Nanocellulose: State of the industry*, December, 2015, Tappi Nano. 10.

[8] Kargarzadeh, H., Mariano, M., Huang, J., Lin, N., Ahmad, I., Dufresne, A., Thomas, S., 2017. Recent developments on nanocellulose reinforced polymer nanocomposites: A review. *Polymer*, Volume 132, 6 December 2017, 368-393. DOI:10.1016/j.polymer.2017.09.043.

[9] El-Saied, H., Basta, A.H., Gobran, R.H., 2004. Research progress in friendly environmental technology for production of cellulose products (bacterial cellulose and its application). *Polym.-Plast. Technol. Eng.*, 43, 797-820. DOI: 10.1081/PPT-120038065.

[10] Brown, E.E., Laborie, M.P.G., 2007. Bioengineering bacterial cellulose/poly(ethylene oxide) nanocomposites. *Biomacromolecules.* 8, 3074-3081. DOI: 10.1021/bm700448x.

[11] Siqueira, G., Bras, J., Dufresne, A., 2010. Cellulosic bionanocomposites: A review of preparation, properties and applications. *Polymers*, 2(4), 728-765, DOI:10.3390/polym2040728.

[12] Golmohammadi, H., Morales-Narváez, E., Naghdi, T., Merkoçi, A., 2017. Nanocellulose in sensing and biosensing. *Chem. Mater.*, 29 (13), 5426-5446. DOI: 10.1021/acs.chemmater.7b01170.

[13] Reese, E.T., 1963. *Advances in enzymic hydrolysis of cellulose and related materials.* Pergamon Press, Oxford, London.

[14] Li, F., Mascheroni, E., Piergiovanni, L., 2015. The potential of nanocellulose in the packaging field: A review. *Packaging Technology and Science*, 28, 475-50808 DOI: 10.1002/pts.2121.

[15] Li, J., Xu, Q.H., Jin, L.Q., 2013. Research development on hydrophobic modification of cellulose nanofibrils. *Adv. Mater. Res.*, 785-786, 440-443. Available from: http://www.scientific.net/AMR.785-786.440.

[16] Robles, N.B., 2014. *Tailoring cellulose nanofibrils for advanced materials*. KTH Royal Institute of Technology, p. 82.

[17] Aulin, C., Ahola, S., Josefsson, P., Nishino, T., Hirose, Y., Osterberg, M., Wagberg, L., 2009. Nanoscale cellulose films with different crystallinities and mesostructures their surface properties and interaction with water. *Langmuir*, 25(13), 7675-7685. DOI:10.1021/la900323n.

[18] Herrick, F.W., Casebier, R.L., Hamilton, J.K., Sandberg, K.R., 1983. Microfibrillated cellulose: Morphology and accessibility, in Applied Polymer Symposia, A. Sarko, (Ed.), 797-813, *Proceedings of the Ninth Cellulose Conference*, Vol. 37, 1983, Wiley, New York City.

[19] Pääkkö, M., Ankerfors, M., Kosonen, H., Nykanen, A., Ahola, S., Osterberg, M., Ruokolainen, J., Laine, J., Larsson, P.T., Ikkala, O., Lindstrom, T., 2007. Enzymatic hydrolysis combined with mechanical shearing and high-pressure homogenization for nanoscale cellulose fibrils and strong gels. *Biomacromolecules*, 8(6), 1934-1941. DOI: 10.1021/bm061215p.

[20] Turbak, A.F., Snyder, F.W., Sandberg, K.R., 1983. Microfibrillated cellulose, a new cellulose product: Properties, uses and commercial potential, in Applied Polymer Symposia, A. Sarko, (Ed.), 815-827, *Proceedings of the Ninth Cellulose Conference*, Vol. 37, 1983, Wiley, New York City, USA.

[21] Li, J., Wei, X., Wang, Q., Chen, J., Chang, G., Kong, L., Su, J., Liu, Y., 2012. Homogeneous isolation of nanocellulose from sugarcane bagasse by high pressure homogenization. *Carbohydr. Polym.*, 90(4), 1609-1613. DOI: 10.1016/j.carbpol.2012.07.038.

[22] Li, J., Xu, Q.H., Jin, L.Q., 2013. Research development on hydrophobic modification of cellulose nanofibrils. *Adv. Mater. Res.*, 785-786, 440-443. Available from: http://www.scientific.net/AMR.785-786.440.

[23] Anderson, S.R., Esposito, D., Gillete, W., Zhu, J.Y., Baxa, U., McNeil, S.E., 2014. Enzymatic preparation of nanocrystalline and microcrystalline cellulose. *Tappi J.*,13(5), 35-42.

[24] https://www.microfluidicscorp.com/microfluidic-technology-equipment/.

[25] http://www.ifoodmac.com/masuko-mkza-10-20j/.

[26] http://analytik.co.uk/product/dispersions-m-110eh-30-microfluidizer/.

[27] http://www.masuko.com/English/product/Masscolloder.html.

[28] Rånby, B.G, Noe, R.W., 1961. *Crystallization of cellulose and cellulose derivatives from dilute solution. I. Growth of single crystals. Journal of Polymer Science*, 51(155), 337-347. DOI:10.1002/pol.1961.1205115520.

[29] Habibi, Y., Lucia, L.A., Rojas, O.J., 2010. Cellulose nanocrystals: chemistry, self-assembly, and applications. *Chem. Rev.*, 110(6), 3479-500. DOI: 10.1021/cr900339w.

[30] Canadian Standards Association, 2014. *Cellulosic Nanomaterials - Test Methods for Characterization* (CSA Z5100-14).

[31] An, X., Cheng, D., Shen, J., Jia, Q., He, Z., Zheng, L., Khan, A., Sun, B., Xiong, B., Ni, Y., 2017. Nanocellulosic materials: research/production activities and applications. *Journal of Bioresources and Bioproducts*, 2(2), 45-49. DOI :10.21967/jbb.v2i2.130.

[32] *Summary of international activities on cellulosic nanomaterials* 2015. www.tappinano.org/media/1096/tc6-world-cnm-activities-summary-july-29-2015.pdf.

[33] *Summary of international activities on cellulosic nanomaterials* 2017. http://www.webctp.com/docs/fckeditor/file/ISO_TC_6__World_CNM_activities_summary_2017-10-20(1).pdf.

[34] http://www.vietworldkitchen.com/blog/2009/04/vietnamese-noodles-101-bun-rice-noodles.html.

[35] http://en.wikipedia.org/wiki/file:afm_innventia_nanocellulose.jpg.www.ontree.ca.

[36] Lagerwall, J.P.F., Schutz, C., Salajkova, M., Noh, J., Park, J.H., Scalia, G, Bergstrom, L., 2014. Cellulose nanocrystal-based materials: from liquid crystal self-assembly and glass formation to multifunctional thin films. *NPG Asia Mater.*, 6, e80, 1-12. DOI : 10.1038/am.2013.69.

[37] Moon, R.J., Martini, A., Nairn, J., Simonsen, J., Youngblood, J., 2011. Cellulose nanomaterials review: structure, properties and nanocomposites. *Chem. Soc. Rev.*, 40, 3941-3994. DOI:10.1039/C0CS00108B.

[38] Bardet, R., Belgacem, N., Bras, J., 2015. Flexibility and color monitoring of cellulose nanocrystal iridescent solid films using anionic or neutral polymers. *ACS Appl. Mater. Interfaces*, 7, 4010-4018. DOI: 10.1021/am506786t.

[39] Beck, S., Bouchard, J., Berry, R., 2011. Controlling the reflection wavelength of iridescent solid films of nanocrystalline cellulose. *Biomacromolecules* 12,167-172. DOI: 10.1021/bm1010905.

[40] Favier, V., Chanzy, H., Cavaille, J.Y., 1995. Polymer nanocomposites reinforced by cellulose whiskers. *Macromolecules*, 28(18), 6365-6367. DOI: 10.1021/ma00122a053.

[41] Favier, V., Canova, G.R., Cavaillé, J.Y., Chanzy, H., Dufresne, A., Gauthier, C., 1995. Nanocomposite materials from latex and cellulose whiskers. *Polymers for Advanced Technologies*, 6(5), 351-355. DOI: 10.1002/pat.1995.220060514.

[42] Xue, L.G., Tabil, L., Panigrahi, S., 2007. Chemical treatments of natural fiber for use in natural fiber-reinforced composites: A review. *Journal of Polymers and the Environment*, 15, 25-33. DOI: 10.1007/s10924-006-0042-3.

[43] Saheb, D.N., Jog, J.P., 1991. Natural fiber polymer composites: A review. *Advances in Polymer Technology*, 18(4), 351-363. DOI: 10.1002/(SICI) 1098-2329(199924)18:4.

[44] Mohanty, A.K., Misra, M., Hinrichsen, G., 2000. Biofibres, biodegradable polymers and biocomposites: An overview. *Macromolecular Materials and Engineering*, 276-277(1), 1-24. DOI: 10.1002/(SICI) 1439-2054(20000301)276:1.

[45] Ngo, T.-D., Ho, T., 2016. Lignocellulosic biomasses, sustainable platform for bio-based materials with polypropylene. Chapter 6 in the book entitled *Polypropylene: properties, uses and benefits*, Paula Garcia, Nova Science Publisher. ISBN: 978-1-53610-008-2.

[46] Moon, R.J., Martini, A., Nairn, J., Simonsen, J., Youngblood, J., 2011. Cellulose nanomaterials review: structure, properties and nanocomposites. *Chem. Soc. Rev.*, 011, 40, 3941-3994. DOI:10. 1039/C0CS00108B.

[47] Zhou, C.-J., Wu, Q.-L., 2012. Recent development in applications of cellulose nanocrystals for advanced polymer-based nanocomposites by novel fabrication strategies. In *Nanocrystals - synthesis, characterization and applications*, Neralla, S., (Chap 6), Intech, Croatia, 103-120. DOI: 10.5772/2560.

[48] Ching, Y.C., Ali, M.E., Abdullah, L., Choo, K., Kuan, Y., Julaihi, S., Chuah, C., Liou, N., 2016. Rheological properties of cellulose nanocrystal embedded polymer composites : A review. *Cellulose*, 23, 1011-1030. DOI : 10.1007/s10570-016-0868-3.

[49] Lima, M.M.D., Borsali, R., 2004. Rodlike cellulose microcrystals: structure, properties, and applications. *Macromol Rapid Commun*, 25(7), 771-787. DOI:10.1002/marc.200300268.

[50] Chen, H., Ding, Y., Lapkin, A., 2009. Rheological behaviour of nanofluids containing tube/rod-like nanoparticles. *Powder Technol.*, 194, 132-141. DOI: 10.1016/j.powtec.2009.03.038.

[51] Reising, A.B., Moon, R.J., Youngblood, J.P., 2012. Effect of particle alignment on mechanical properties of neat cellulose nanocrystal films. *J. Sci. Technol. For. Prod. Process*. 2, 32-41.

[52] Li, F., Mascheroni, E., Piergiovanni, L., 2015. The potential of nanocellulose in the packaging field: A review. *Packag. Technol. Sci.*, 28, 475-508. DOI: 10.1002/pts.2121.

[53] Skatkin, J.A., Wegner, T.H, Bilek E.M.(Ted), Cowie, J., 2014. Market projections of cellulose nanomaterial-enabled products - Part 1: Applications. *Tappi Journal*, 13(5), 9-16.

[54] Cowie, J., Bilek, E.M.(Ted), Wegner, T.H., Shatkin, J.A., 2014. Market projections of cellulose nanomaterial-enable products-Part 2: Volume estimation, *Tappi journal*, 13(6), 57-69.

[55] Miller, J., 2014. *Nanocellulose: technology applications, and markets*, RISI.

In: Cellulose to Nanocelluloses
Editor: Tri-Dung Ngo
ISBN: 978-1-53615-057-5
© 2019 Nova Science Publishers, Inc.

Chapter 3

SYNTHETIC METHODS TO PRODUCE CARBOXYLATED CELLULOSE NANOCRYSTALS

Alfred C. W. Leung[1,*] and Edmond Lam[1]
[1]Aquatic and Crop Resource Development Research Center,
National Research Council of Canada,
Montreal, QC, Canada

ABSTRACT

Cellulose nanocrystals (CNCs) are typically synthesized by hydrolysis of cellulosic fibers using strong acids. In the occasions when high surface charge or reactive sites on the CNC surface is desired, a popular strategy is to oxidize CNCs to obtain carboxylic acid groups. This review chapter discusses the various synthetic procedures to prepare carboxylated cellulose nanocrystals (cCNCs). The common goal of these procedures is to develop a simple high yielding process that oxidizes the surface primary hydroxyl groups with high conversion rate, while leaving the secondary hydroxyl groups and the crystallinity intact.

[*] Corresponding Author Email: alfred.leung@cnrc-nrc.gc.ca.

Keywords: carboxylated cellulose nanocrystals, nanocrystalline cellulose, nanocellulose

INTRODUCTION

Cellulose, a semi-crystalline polymer of 1,4-linked β-D-glucopyranose, consists of regions that are amorphous and crystalline. The crystalline regions have a highly compact and ordered structure which provides the polymer with excellent mechanical properties and chemical resistance. The amorphous regions are disordered and flexible, and they are more susceptible to chemical attack.

CNCs can be prepared by removing the amorphous regions, yielding nanoparticles of crystalline cellulose. A number of procedures have been developed for the synthesis of CNCs, and the most commonly employed procedure uses highly concentrated sulfuric acid for the controlled hydrolysis of cellulose [1]. Other methods to isolate CNCs include the use of enzymes, mechanical treatments, oxidizing agents or a combination thereof [2].

Depending on the application of the CNCs, additional surface chemical groups such as the sulfate ester or carboxyl are often desired. The presence of surface reactive groups such as the carboxylic groups enable numerous routes to modify the surface chemistry of the CNC, especially for the attachment of functional molecules such as polymers, enzymes, fluorophores or metallic nanoparticles [3]. Examples of potential applications for which the use of cCNC may be advantageous over sulfated-CNC are as additives for flexible packaging films where superior transparency and oxygen barrier properties were reported, and as "nanopaper" substrates for printed electonics where higher transparency and tensile strength were also observed [4].

This review chapter provides an overview of the various literature methods to synthesize cCNCs and a description of some of their unique features.

SYNTHESIS OF CCNCS

Catalytic oxidation using (2,2,6,6-tetramethylpiperidin-1-yl)oxyl (TEMPO), a commercially available nitroxyl radical, is an efficient procedure for the oxidation of alcohols to ketones, aldehydes, and carboxyl groups under relatively mild conditions [5]. Starting with CNC hydrolyzed by HCl that does not contain any surface groups, Montanari and coworkers applied TEMPO oxidation to the CNC to produce cCNC. This oxidation method uses a mixture of sodium hypochlorite (NaClO), sodium bromide, and TEMPO catalyst to regioselectively oxidize the C6 primary hydroxyl groups, Scheme 1. The concentration of carboxyl groups can be varied by changing the NaClO concentration, with the maximum degree of oxidation (DO) value corresponding to the number of available surface hydroxyl groups. Nanofibers of very high aspect ratio (μm in length) can be obtained by TEMPO oxidation of native cellulosics followed by a mechanical disintegration procedure such as sonication or high pressure homogenization [6].

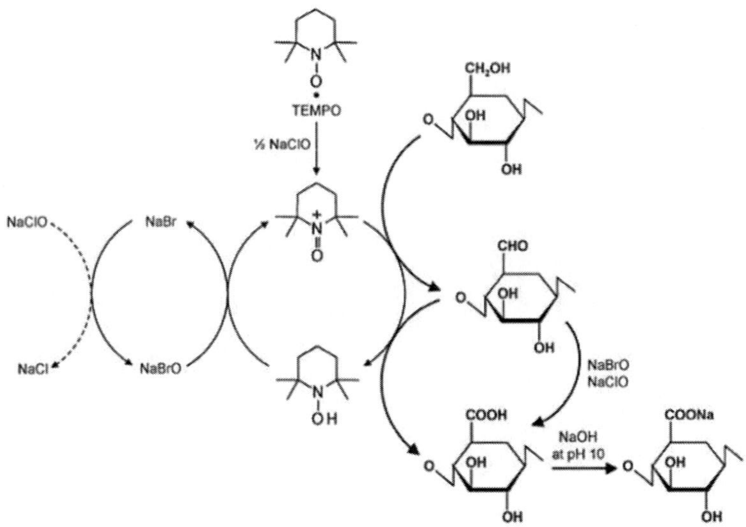

Scheme 1. Mechanism of TEMPO catalysed oxidation of cellulose at the C6 methylhydroxyl group to the carboxylate group. (Reproduced from Ref. [5] with permission of The Royal Society of Chemistry).

Due to the high aspect ratio of TEMPO oxidized nanofibers, solution viscosity is higher and gelation occurs at a lower concentration. Upon drying, flexible and transparent films, rather than brittle birefringent films as in the case of CNCs, can be produced. Similar to sulfated CNC, TEMPO oxidized cCNC exhibited a chiral nematic liquid crystalline phase [7]. The critical concentration of liquid crystals formation decreases as the aspect ratio of the nanocrystals increases.

Leung and coworkers reported a one-step procedure to prepare cCNCs by heating cellulosic materials at 60 °C in 1M ammonium persulfate (APS). It normally requires 16 h for cellulosic fibers such as hemp, flax and triticale, but the optimal reaction time varies with different substrates. In a one-step reaction, APS is capable of removing lignin, hemicellulose, pectin, and other plant contents, generating high quality cCNCs. AFM and TEM micrographs illustrate the production of highly uniform cCNCs, Figure 1.

The cross-section dimensions of cCNCs prepared from APS treatment of hemp, flax and triticale are centered at 3 – 6 nm, consistent with elementary fibrils (AFM, 3 – 7 nm in diameter) of cellulose. As shown in Table 1, cCNCs samples prepared from different cellulosics showed a similar mean particle length and length polydispersity.

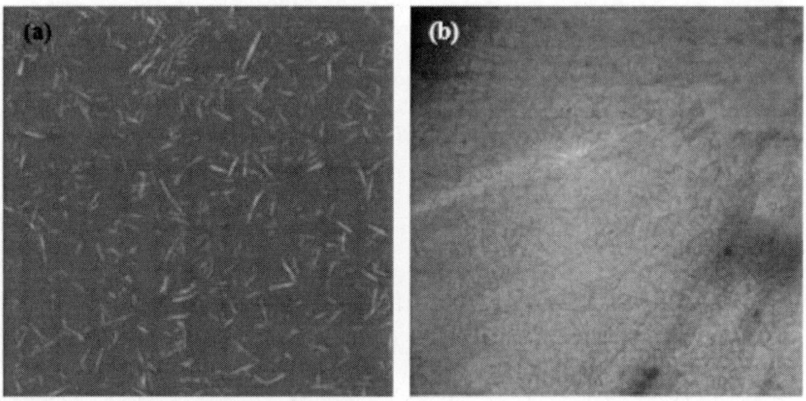

Figure 1. (a) AFM micrograph (5 μm) of cCNCs produced from APS treatment of hemp fibers and (b) Low Voltage Transmission Electron Microscopy (LVTEM) micrographs of cCNCs from hemp fibers.

Table 1. Yield, degree of oxidation, crystallinity and dimensions of cCNCs from various sources. CRI are estimated using PXRD. Length and diameter are reported at 95% confidence interval (S. M. denotes starting materials)

Cellulose type	Yield (%)	Degree of oxidation	CRI (%) S. M.	CRI (%) cCNC	Length (std. dev. (SD), nm)	Diameter (std. dev. (SD), nm)
Flax	28	0.18	69	75	144 ± 5 (n = 138, SD = 53)	3.8 ± 0.1 (n = 1653, SD = 1.9)
Flax shives	22	0.19	48	64	296 ± 16 (n = 445, SD =194)	5.1 ± 0.1 (n = 1350, SD = 1.7)
Hemp	36	0.17	56	73	148 ± 3 (n = 292, SD = 52)	5.8 ± 0.1 (n = 906, SD = 1.8)
Triticale	31	0.11	61	73	134 ± 5 (n = 118, SD = 54)	4.2 ± 0.1 (n = 266, SD = 1.8)
MCC	65	0.19	79	83	128 ± 4 (n = 100, SD = 38)	5.5 ± 0.1 (n = 2291, SD = 1.5)
Whatman CF1	81	0.12	91	91	121 ± 3 (n = 129, SD = 39)	6.7 ± 0.3 (n = 207, SD = 2.2)
Wood pulp	36	0.19	73	81	124 ± 6 (n = 200, SD = 41)	6.0 ± 0.2 (n = 253, SD = 1.7)
Bacterial cellulose	14	0.14	59	70	88 ± 5 (n = 500, SD = 55)	6.5 ± 0.2 (n = 497, SD = 2.3)

This was a very important finding since the uniformity, small diameter, and high aspect ratio of the nanocrystals are critical for their intended applications such as nanofillers or for drug delivery. The reaction mixture was analyzed using Raman spectroscopy, which suggests that APS was rapidly decomposed to form $SO_4^{\cdot-}$ and HSO_4^-, which in turn co-hydrolyzed the 1,4-β bond of the cellulose chain in the amorphous region resulting in the formation of cCNCs. Together with hydrogen peroxide, these two anions were capable of oxidizing the $-CH_2OH$ in the C6 position to form carboxylic acid groups on the CNC surface. The major by-products at the end of the reaction were ammonium sulfate and sulfuric acid, present in a molar ratio of approximately 4:6, respectively. HPLC analysis of the spent liquor confirmed the absence of furfural or 2-hydroxymethylfurfural (HMF), two hazardous chemicals which are often encountered when carbohydrates are hydrolyzed using strong acids. Under similar reaction conditions, highly uniform spherical cCNC with an average diameter of 35 nm and surface carboxylic acid group with DO up to 0.14 can be produced by using Lyocell, a cellulose II substrate, as the starting material [8]. Rod shape cCNC produced by APS exhibits chiral nematic liquid crystalline phases that are similar to CNC prepared by H_2SO_4 hydrolysis. However, it was noted that these cCNC required additional time to form an ordered phase compared to CNC containing sulfate half ester groups, possibly due to the weak acidic nature of the carboxylic acid groups [9].

Hydrothermal system using high pressure and high temperature is a potential method to speed up the reaction time of CNC extraction. At 110°C using 6M HCl, CNC can be extracted from microcrystalline cellulose (MCC) at 93% yield in 3 h [10]. Cheng and coworkers reported that the addition of HNO_3 as an oxidant to HCl under similar hydrothermal condition yielded rod shaped cCNC, with a significant reduction in the HCl concentration required for the reaction from 6M to 4M HCl [11]. Size analyses of the cCNCs were performed using FE-SEM and DLS. It was observed that cCNC size decreased with increasing HNO_3 concentration, suggesting that the addition of HNO_3 accelerates the hydrolysis of amorphous cellulose. The authors reported that the optimal HCl:HNO_3 ratio of 7:3 resulted in the smallest nanocrystals dimensions, DO, and yield of (176 ± 6) nm × (11 ± 2) nm (L × W), 0.12, and 73%, respectively. The CRI of cCNC increased slightly from 84% for MCC to 92% in cCNC. The increase in CRI is indicative that carboxylation occurred on the surface and the main crystal structure was not affected significantly.

cCNC can be prepared by hydrolysis using dicarboxylic acids such as oxalic acid and maleic acid [12]. At room temperature, these organic acids can be crystallized in water, facilitating the recovery of the unspent reagents. At elevated temperatures, these organic acids are soluble, and carboxyl functionalized cCNC can be synthesized through a Fischer-Speier esterification process. The optimal conditions that produce cCNC were achieved using 70% oxalic acid at 100 °C for 1 h. The cCNC yield was relatively low (up to 25%) due to the weak nature of oxalic acid and maleic acid compares to mineral acid such as H_2SO_4. However, the reaction waste can be minimized by recovering the cellulosic residue from the first reaction and mechanically fibrillated to produce cellulose nanofibrils (CNF), Scheme 2.

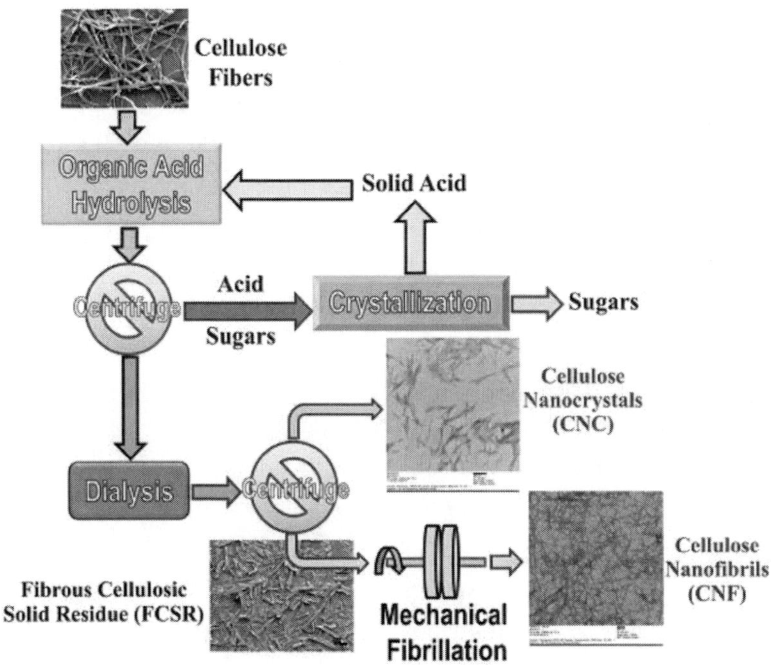

Scheme 2. Flow diagram for integrated production of cCNC and CNF and the recycling of the spent organic acid. (Reproduced from Ref. [11a] with permission of The Royal Society of Chemistry).

cCNC dimensions are determined to range from 270 – 370 nm using AFM, and they are relatively longer than cCNC produced using conventional acid hydrolysis methods. The larger crystal size and the low CRI (81%) may be indicative of incomplete removal of amorphous cellulose. A comparison of different cellulosic starting material of bleached eucalyptus pulp (BEP), spruce dissolving pulp (SDP) and cotton based filter paper (CFP) revealed a large variation in particle size of approximately 500 nm, 100 – 200 nm and 2 μm, respectively.

A commercially available oxidant under the trade name Oxone® containing sulfates ($KHSO_5$, $KHSO_4$, and K_2SO_4) was reported by Ruan and co-workers for the production of cCNC [13]. A major advantage of using Oxone® is its relatively low cost and low toxicity. The cCNC produced using this method contains a small amount of aldehyde and ketone groups, which can be reduced by the addition of NaBr, while increasing the carboxylic acid content by 10 – 20%. Similar to cCNC produced by APS and TEMPO, oxidation by Oxone® preferentially targets the C6 and possibly to some extent the C1 on the reducing end group. However, a relatively low carboxyl content of 0.45 mmol/g was reported for the optimized reaction conditions. cCNC produced using this procedure may not be suitable for applications requiring a high surface charge on the nanocrystals.

Yu and coworkers reported the use of a mixed citric acid and hydrochloric acid solution to produce cCNC in one step from MCC [14]. cCNC with carboxyl content of up to 1.4 mmol/g and average dimensions of (220 ± 20) nm × (16 ± 3) nm (L × D, FE-SEM) can be prepared by heating a 2 wt% solution of MCC at 80 °C for 4 h for a yield of 82.5%. cCNC yield can be increased to 90% by reducing the heating time to 2 h, but the carboxyl content of the resulting nanocrystals are lower (1.15 mmol/g) and the suspension stability is negatively affected. The authors also reported that the hydrophilicity of cCNC increased with carboxyl content as measured by the static contact angle, and demonstrated a potential application of cCNC as flocculants for textile effluents containing organic dyes.

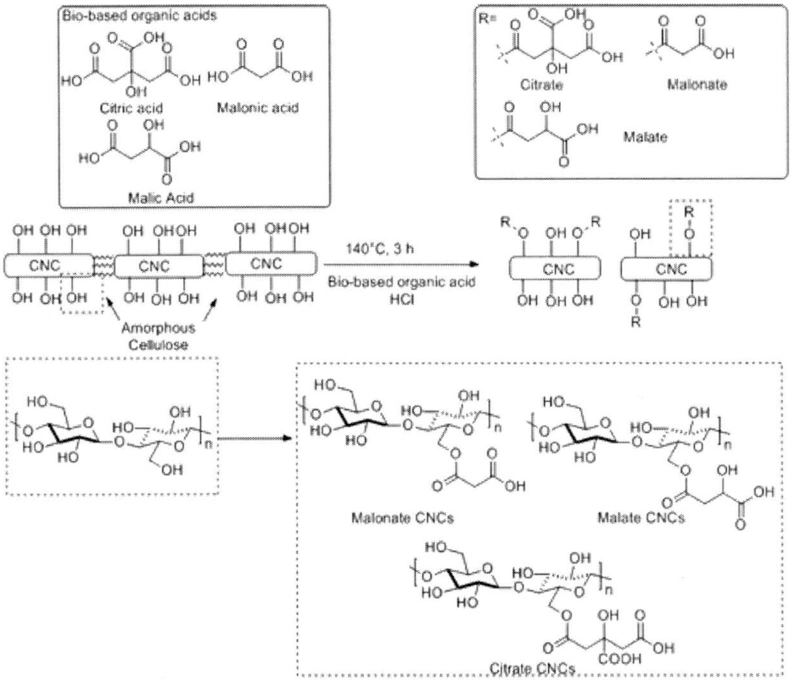

Scheme 3. cCNC produced by one pot acid hydrolysis and Fischer esterification. (Reprint with permission from ref. [14]. Copyright 2015 American Chemical Society).

A concurrent acid-catalyzed hydrolysis and Fischer esterification process was developed by Spinella and coworkers by combining HCl and organic acid such as malonic, malic, and citric acid, Scheme 3 [15]. Using only the organic acids, the cCNC yield was below 10%, consistent with other processes using neat organic acids. The addition of up to 0.1 M HCl significantly improved product yield up to 25%, but higher HCl concentration at or above 0.2M led to discoloration of the product. The discoloration was due to the hydrolysis of the cellulose's crystalline domains, resulting in glucose decomposition products such as hydroxymethyl-furfural and humic acids. TEM analysis shows rod shape particles with dimensions of ~ 220 × 12 nm (L × W), and PXRD indicates the CRI are 75 – 78%, consistent with other CNC made from ramie cellulose. The carboxyl content was highest for cCNC produced from citric acid (1.9 mmol/g), followed by cCNC produced from malonate acid (1.6 mmol/g), and malate acid (1.1 mmol/g). These results indicate that citric acid is a superior reagent for the generation of cCNC with high surface carboxyl content.

CONCLUSION

Owing to their high surface charge density and ease of modification, cCNC are often used as a scaffold to build more complex structures and composites. While the process to produce CNC using H_2SO_4 hydrolysis has reached commercial pilot scale, scale-up and production of cCNC are met with several challenges such as the need for multi-step synthesis and the use of toxic catalysts. Current research is focused on the development of a simple one-step reaction, and the recovery and recycling of reaction waste and unspent reagents to afford an efficient, cost effective "green" process.

REFERENCES

[1] Revol, J.-F., Bradford, H., Giasson, J., Marchessault, R. H. and Gray, D. G. (1992). Helicoidal self-ordering of cellulose microfibrils in aqueous suspension. *International Journal of Biological Macromolecules*, 14 (3): 170-172.

[2] Dieter, K., Kramer, F., Moritz, S., Lindström, T., Ankerfors, M., Gray, D. and Dorris, A. (2011). Nanocelluloses: a new family of nature-based materials. *Angewandte Chemie International Edition,* 50 (24): 5438-5466.

[3] Youssef, H., Lucia, L. A. and Rojas., O. J. (2010). Cellulose nanocrystals: chemistry, self-assembly, and applications. *Chemical Reviews,* 110(6): 3479-3500.

[4] (a) Mascheroni, E., Rampazzo, R., Ortenzi, M. A., Piva, G., Bonetti, S. and Piergiovanni, L. (2016) Comparison of cellulose nanocrystals obtained by sulfuric acid hydrolysis and ammonium persulfate, to be

used as coating on flexible food-packaging materials. *Cellulose*, 23(1): 779-793. (b) Sun, X., Wu, Q., Ren, S. and Lei, T. (2015) Comparison of highly transparent all-cellulose nanopaper prepared using sulfuric acid and TEMPO-mediated oxidation methods. *Cellulose*, 22(2): 1123-1133.

[5] Suzelei, M., Roumani, M., Heux, L. and Vignon, M. R. (2005). Topochemistry of carboxylated cellulose nanocrystals resulting from TEMPO-mediated oxidation. *Macromolecules,* 38(5): 1665-1671.

[6] Akira, I., Saito, T. and Fukuzumi, H. (2011). TEMPO-oxidized cellulose nanofibers. *Nanoscale,* 3(1): 71-85.

[7] Jintao, H., Liu, S., Li, L. and Piao, G. (2017). Lyotropic liquid crystal behavior of carboxylated cellulose nanocrystals. *Carbohydrate Polymers*, 164: 364-369.

[8] Miao, C., Qin, Z., Liu, Y., Qin, Y., Li, T., Chen, L. and Zhu, M. (2014). Efficient extraction of carboxylated spherical cellulose nanocrystals with narrow distribution through hydrolysis of lyocell fibers by using ammonium persulfate as an oxidant. *Journal of Materials Chemistry A,* 2(1): 251-258.

[9] Guerrero, C., Carlos F. and Gray, D. G. (2014). Chiral nematic phase formation by aqueous suspensions of cellulose nanocrystals prepared by oxidation with ammonium persulfate. *Cellulose*, 21(4): 2567- 2577.

[10] Houyong, Y., Qin, Z., Liang, B., Liu, N., Zhou, Z. and Chen, L. (2013). Facile extraction of thermally stable cellulose nanocrystals with a high yield of 93% through hydrochloric acid hydrolysis under hydrothermal conditions. *Journal of Materials Chemistry A*, 1(12): 3938-3944.

[11] Miao, C., Qin, Z., Chen, Y., Liu, J. and Ren, Z. (2017). Facile one-step extraction and oxidative carboxylation of cellulose nanocrystals through hydrothermal reaction by using mixed inorganic acids. *Cellulose*, 24(8): 3243-3254.

[12] (a) Liheng, C., Zhu, J. Y., Baez, C., Kitin, P. and Elder, T. (2016) Highly thermal-stable and functional cellulose nanocrystals and nanofibrils produced using fully recyclable organic acids. *Green Chemistry*, 18(13): 3835-3843. (b) Wang, R., Chen, L., Zhu, J. Y. and Yang, R. (2017). Tailored and integrated production of carboxylated

cellulose nanocrystals (CNC) with nanofibrils (CNF) through maleic acid hydrolysis. *Chem. NanoMat.,* 3(5): 328-335. (c) Chao, J., Chen, L., Shao, Z., Agarwal, U. P., Hu, L. and Zhu, J. Y. (2017). Using a fully recyclable dicarboxylic acid for producing dispersible and thermally stable cellulose nanomaterials from different cellulosic sources. *Cellulose*, 24(6): 2483-2498.

[13] Chang-Qing, R., Strømme, M., Mihranyan, A. and Lindh, J. (2017). Favored surface-limited oxidation of cellulose with Oxone® in water. *RSC Advances*, 7(64): 40600-40607.

[14] Hou-Yong, Y., Zhang, D. Z., Lu, F. F. and Yao, J. (2016). New approach for single-step extraction of carboxylated cellulose nanocrystals for their use as adsorbents and flocculants. *ACS Sustainable Chemistry & Engineering,* 4(5): 2632-2643.

[15] Stephen, S., Maiorana, A., Qian, Q., Dawson, N. J., Hepworth, V., McCallum, S. A., Ganesh, M., Singer, K. D. and Gross, R. A. (2016). Concurrent cellulose hydrolysis and esterification to prepare a surface-modified cellulose nanocrystal decorated with carboxylic acid moieties. *ACS Sustainable Chemistry & Engineering*, 4(3): 1538-1550.

In: Cellulose to Nanocelluloses
Editor: Tri-Dung Ngo
ISBN: 978-1-53615-057-5
© 2019 Nova Science Publishers, Inc.

Chapter 4

CELLULOSE NANOFIBER, PROPERTIES AND APPLICATIONS

Le Van Hai[1,2,*], Lindong Zhai[1], Hyun Chan Kim[1], Jung Woong Kim[1], Ruth M. Muthoka[1], Abdullahil Kafy[1] and Jaehwan Kim[1,*]

[1]CRC for Nanocellulose Future Composite, Inha University, Incheon City, Republic of Korea

[2]Department of Pulp and Paper Technology, Phutho College of Industry and Trade, Phutho, Vietnam

ABSTRACT

Cellulose is one of the most abundant natural material on earth; mainly from plants. Being a fibrous material, it is the basis of man-made fibers finding its way in many application areas such as construction, building, food, cosmetics, papers, packaging and so on. Cellulose fibers can be extracted from different cellulose sources. Cellulose nanofiber (CNF) is a new type of material that is at least less than 100 nm in one dimension. In general, the extraction of cellulose nanofiber can be through mechanical

* Corresponding Author Email: levanhai121978@gmail.com and jaehwan@inha.ac.kr.

means, enzymatic, chemical, and growing bacterial cellulose. Mechanical extraction is usually done by supermasscolloider, high-pressure homogenizer and others. The cellulose nanofiber formed by mechanical methods are known as nanofibrillated cellulose (NFC), sometimes referred to as microfibrillated cellulose (MFC). In addition, several research groups use a common mechanical method, aqueous counter collision (ACC). Fibers produced from the ACC method are very long in length and have a small width. Chemical method of extracting CNF involve the use of acids, mainly hydrochloric acid, sulphuric acid, and nitric acid. This method produces nanocrystals cellulose (NCC) which are often referred to as cellulose nanowhisker or nanocrystals cellulose. Another approach entails TEMPO-oxidation method which is capable of producing cellulose nanofiber with a very high aspect ratio. In this chapter, in the case of nanofibrillated cellulose, nanocrystals cellulose or nanofibril cellulose we will refer to as cellulose nanofiber. The extracted cellulose nanofiber is used for various purposes such as reinforcement, coating, lubrication, in composite materials, and smart materials among others. This chapter focuses on different isolation methods of cellulose nanofiber as well as detailed discussion on its properties such as mechanical properties, thermal degradation properties, and crystallinity index. In addition, several potential applications of cellulose nanofiber such as in making composites, smart materials, food packaging, coating, drug delivery, and other are also going to be discussed.

Keywords: nanofibrillated cellulose, nanocrystal cellulose, cellulose nanofiber, all-cellulose nanocomposite, smarter materials

INTRODUCTION

Cellulose is a polymer whose formula is $(C_6H_{10}O_5)_n$. It is a polysaccharide consisting of a long linear chain of β (1-4) glycosidic bonds. The structure of cellulose was first named by Anselme Payen in 1938 [1]. Cellulose is a green material; bio-degradable, environmentally friendly, carbon neutral and abundantly available on earth. Cellulose fiber can be extracted from wood and non-wood materials through the process of pulping. Cellulose is used for paper making, packaging, and preparation of cellulose derivatives (rayon, cellophane…), cellulose composites, and smart materials. Cellulose nanofiber, a new material of concern, is extracted by chemical, mechanical, physical and growing bacterial as indicated by research groups [2-6]. To consider cellulose as cellulose nanofiber, at least one dimension has to be less than 100 nm. Cellulose nanofiber can be extracted from different cellulose sources such as wood (hardwood, softwood), non-wood (red algae, cotton linter, tunicate, grasses, mangosteen, Sugarcane) [7-12]. Most of the cellulose nanofiber is extracted from bleached pulp. However, several research groups used non-bleached, thermomechanical pulp to produce cellulose nanofiber as reported by Okita et al. 2009 [13]. Cellulose nanofiber has gained much attention due to its light-weight, high tensile strength, Young's modulus and abundance. According to Lahiji et al. 2008 [14] and Moon et al. 2011 [7] the mechanical properties of cellulose nanofiber are 5-7.5 GPa and 57-145 GPa for tensile strength and Young's modulus respectively.

In general, wood and nonwood are the native resources of cellulose nanofibers. In recent years, cellulose and cellulose nanofibers were also produced from recycled fiber sources as an attempt to save the environment. Several research groups use waste papers (corrugated paper, Office waste paper) to extract nanocellulose [9, 15-18]. Wang et al. 2013 [16] isolated nanofibrillated cellulose using the mechanical method. Danial et al. 2015 [17] used waste paper to produce cellulose nanocrystal and evaluated its properties. Takagi et al. 2013 [15] produced nanocrystal cellulose and used it for developing bio-degradable composites. Hai et al. 2017 [9] used office waste paper to produce nanocellulose by TEMPO-oxidation treatment and made a comparison to native cellulose sources in terms of different aspects such as morphology, XRD, TGA. In addition, Hai et al. 2017 [9] used cellulose nanofiber from office waste paper to reinforce in all-cellulose composite.

To be considered a nanosizefiber; at least one dimension must be smaller than 100 nm. It is, in general, most publications mentioned the width of cellulose nanofiber as 2-20 nm for cellulose nanocrystals, 2-10 nm for cellulose microfibrils and 10-40 nm for nanofibrillated cellulose [7, 19]. The length of cellulose nanofiber depends on the fabrication methods. The length of CNC is usually less than 1 μm; however, it is also depended on the sources of cellulose. From a study conducted by Moon et al. 2011 [7], CNC extracted from Tunicate, Bamboo, and OPB had a length longer than 1 μm. Nanofibrillated cellulose or sometimes called microfibrillated cellulose has a long length from several hundred nm to several μm in length. To define the cellulose morphology, different methods can be used. Transmission electron microscopy (TEM), Field emission scanning electron microscopy (FE-SEM), Atomic force microscopy (AFM), Dynamic Light scattering (DLS) and nano particles analyzer are used to analyse the width and length of cellulose nanofibers. In general, the length of nanofibrillated cellulose is very difficult to determine. Some publications mentioned that the length of cellulose nanofiber could be measured by DLS method. The reported length of the nanofibrillated or microfibrillated cellulose is generally several microns. Only a few articles are concerned by the methods to determining the length of nanofibrillated cellulose [20]. The length of nanofibrillated cellulose was determined by two different approaches using FE-SEM and image J [20]. However, determining the length of nanofibrillated cellulose by FE-SEM, TEM or AFM is time-consuming. A detailed procedure and evaluation method are introduced for length determination.

For evaluation of thermal stability of cellulose nanofiber, thermogravimetric analysis is usually used. Many research papers have shown that the thermal stability of nanocellulose has responds differently. Nanocrystals cellulose, in general, has lower thermal stability compared to native cellulose sources. However, nanocellulose extracted by mechanical treatment such as using blender and grinder, usually show similar thermal stability to that of native cellulose sources. Many reports mentioned that the thermal stability of cellulose nanofiber varies from around 180°C to 400°C. Also, nanocellulose may have different response in terms of thermal stability. In the higher temperature range around 400°C, nanocellulose fiber was observed to be more thermally stable than that of the cellulose sources. However, it has not been of much concern, with few discussion and results reported to back upon that point. Therefore, this can be an interest for further investigation and study.

The crystallinity index of cellulose and nanocellulose materials is important for mechanical properties of materials. Thus, for cellulose, cellulose nanofibers, the crystallinity index is always a major concern. The crystallinity index is evaluated by X-ray diffraction method (XRD). Native cellulose and nanocellulose have different reflection in crystallinity index. Several researches mentioned the CrI to be higher for cellulose nanocrystals compared to that of cellulose sources but other researchers disputed with opposite results. In the case of nanofibrillated cellulose, the crystallinity index has a very low effect. The crystallinity index of cellulose nanofiber has already been reported in a large number of publications. To evaluate the crystallinity index, the powder form or the thin film forms of cellulose nanofibers are generally used. There are also many other approaches in calculating crystallinity index. Commonly, the Segal method is a very popular method and easy to use. The Segal method calculates the crystallinity index of cellulose and nanocrystalline cellulose from the $2\theta =$ 21 and 17.8 degrees [21].

Cellulose nanofiber can also be used as a reinforcing material. For composite applications, nanocellulose has attracted a lot of attention due to its light-weight, high mechanical properties and environmental friendliness. Nanocellulose fiber has been used for different type of composites such as Polymer lattices [22], POE [23], PE, LDPE, PVC, PVA, and PLA [24-25]. Tagaki et al. 2013 [15] extracted cellulose nanofiber from waste papers and reinforced it in a bio-degradable composite. They mentioned the improvement of mechanical properties of the resulting composite. In addition, several research groups used cellulose reinforcement in dissolved cellulose matrix, which formed all-cellulose biocomposites [8-9, 26].

Furthermore, nanocellulose plays important role as a functional material according to reports [27-30], with applications in heavy metals adsorption [31], drug delivery [32], nanocellulose-PEG bionanocomposite wound dressings [33], composite membrane using cellulose nanofiber [34-35], and smart materials [36]. In addition, cellulose nanocrystals are being used for supercapacitor materials [37] and are being considered for use in multifunctional thin films [38], as electro-conductive composites [39], as well as in organic solar cells [40].

Cellulose nanofibers have also been studied for different applications such as coating, lubricants, colors, and others [36]. Some research groups mentioned cellulose nanofiber was used for paper coating [41-42]. Using nanocellulose fiber for coating reduces the surface porosity while air resistance, strength, and stiffness were improved [41-43].

METHODS

Nanocrystals Cellulose

Nanocrystals cellulose is isolated using sulfuric acid, hydrochloric acid or others acids. In 1949 Ranby mentioned cellulose micelles in his research. In the same period in 1950, he mentioned the cellulose crystals in another report and another group Mukherjee and Woods 1953 [44] mentioned it in their work as well. The research on cellulose nanocrystals had not gained much attention until recent years. There are different recipes for producing nanocrystals cellulose. In general, to produce nanocrystal cellulose using sulfuric acid, some factors such as acid concentration, isolation time, and temperature are critical in the overall properties of nanocrystals cellulose. Some of the common conditions in isolating nanocrystals are concentrated sulfuric acid 65% by weight at a temperature range of 50°C to 70°CoC, and a time range from very short of about 15 minutes to several hours. The samples are then diluted with around 10-fold of water, then washed several times followed by centrifugation. From research groups [45-46], they used 64%wt sulfuric acid at 45°C for 45 minutes to extract CNC. Revol et al. 1992 [47], used sulfuric acid 65% at 70°C for 30 minutes to isolate CNC. Gaspar et al. 2014 [48], produced nanocrystals cellulose at 45°C in 130 minutes. Santos et al. 2014 [49] prepared cellulose nanocrystal using sulfuric acid 60%wt at 45°Cfor 60 minutes. Fahma et al. 2010 [50] used 64% sulfuric acid at 45°C for a time range of 15 to 90 minutes. Nanocrystals cellulose from different cellulose sources was hydrolyzed using hydrochloric acid [51-53]. According to Winter et al. 2010 [52], bacterial cellulose nanocellulose were isolated using 2.5M HCl at 70°C for 2 hours reaction time. Also, they reported the yield of NCC as 78%, with a width 18 nm and length about 908 nm. George et al. 2014 [53], isolated nanocrystals cellulose using 500 ml of 4N HCl, 100 g of cellulose, and 4hr in boiling conditions. Froma study done by Yu et al. 2013 [54], nanocrystals cellulose was extracted from commercial MCCs. Gaspar et al. 2014 [55], prepared nanocrystals cellulose from cotton linter source.

NCC can also be prepared using Phosphotungstic acid [56]. NCC was isolated using 40ml ($H_3PW_{14}O_{40}$) 0.5 g cellulose, 50 to 80% HPW at a temperature of 90°C under mechanical stirring for 15-30hr [56]. According to Liu et al. 2014 [56], the cellulose nanofiber has a width from ranging15-25 nm and length of 600-800 nm. Also the yield of NCC can be improved from around 30% to 60% after 30 hr reaction with 75%w/w of HPW.

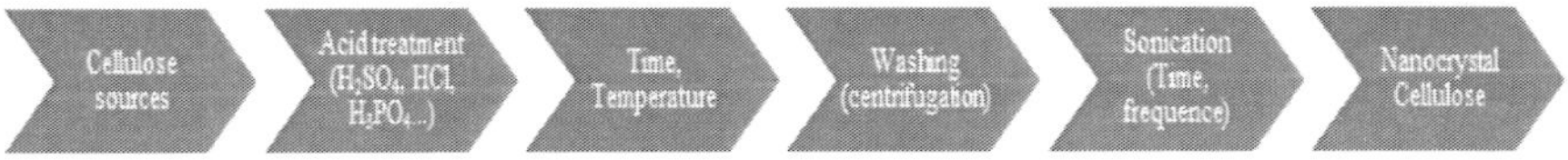

Figure 1. Illustration nanocrystal cellulose isolation process.

TEMPO-Oxidized Cellulose Nanofiber

TEMPO-oxidation cellulose was first discovered by Isogai & Kato 1998 [5]. The TEMPO-oxidized cellulose nanofiber was then extracted as indicated [57-58]. According to Satito et al. [57] the TEMPO-oxidized cellulose nanofiber is very long in length, with a small width. It is flexible and retains its crystallinity. Many research papers have claims that the most convenient source for extracting cellulose nanofiber is never dried pulp compared to dried one. According to research groups, TEMPO-oxidation cellulose nanofiber is easy to isolate and need a simple device for extraction. Since then, there has been a huge increment in the number of published papers concerning TEMPO-oxidation of cellulose nanofiber, and its applications. For TEMPO-oxidation, the chemicals involved include TEMPO, NaCl, NaBr, NaOH, and HCl. Different research groups adjusted the chemical dosage of the above chemicals to achieve their intended properties. In one of the early papers on TEMPO-oxidation, Isogai, and Kato 1998 [5], a variation of oxidation of cellulose was achieved by changing the chemical dosage. Isogai and Kaito1998 [5], TEMPO-oxidation of cellulose was carried out in different conditions and materials. In their first attempted

TEMPO-oxidation, they used 7-20 ml of 11% NaClO, for 1.5 h to 9.5 h, at a temperature of 4°C and 27°C, and applied 0.48 g of TEMPO. Since then, many papers on TEMPO-oxidation method of isolating nanofibrils cellulose have been published. Saito et al. 2006 [57], used never dried cotton and never dried bleached sulfite pulp from Pinus pinaster to isolate cellulose nanofibers. The dosage of the NaClO was 0 to 6.3 mmol/g and reaction time was set from 5 to 120 minutes. After washing, the TEMPO-oxidized cellulose nanofiber was mechanically treated using Waring Blender. The author reported that after 3 successive TEMPO-oxidation treatment, the cellulose nanofiber could reach 99%. Still in another paper, Saito et al. 2007 [58], produced TEMPO-oxidized nanocellulose fiber from TEMPO (0.016 g), NaBr (0.1 g), NaClO (1.3 to 5.0 mmol), at a pH 10 adjusted using 0.5M NaOH under a mechanical stirrer at 500 rpm in room temperature. In that paper, the authors reported that the individual cellulose nanofiber had a width of 3-4 nm and was very long length. Okita et al. 2009 [13], produced cellulose nanofiber from never dried softwood thermomechanical pulp (TMP). 5 g of TMP was treated with TEMPO (0.0625 g), NaBr (0.625 g) and NaClO (0.8-26 mmol), at a pH 10. Also, recently Hai et al. 2017 [9] prepared cellulose nanofiber by employing a combined method of TEMPO-oxidation treatment followed by aqueous counter collision (ACC). The resulting cellulose nanofiber was then used in green all-cellulose composite [9], along cellulose fiber with strong and tough characteristics Kafy et al. 2017 [59].

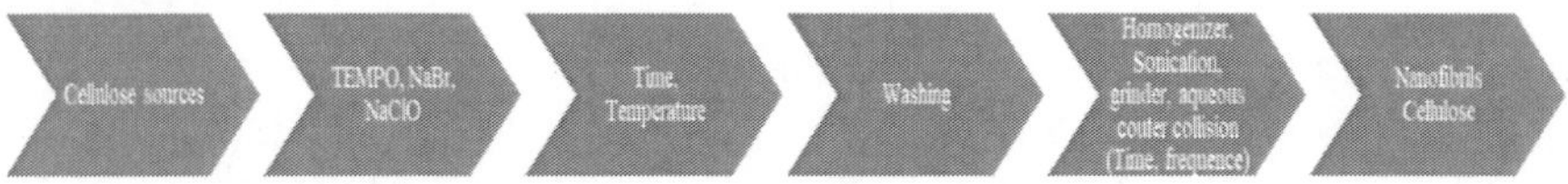

Figure 2. Illustration nanofibrils cellulose isolation process.

Nanofibrillated Cellulose

To produce nanofibrillated cellulose, there are different approaches that can be employed. Such methods include the use of supermasscolloider, a high-pressure homogenizer, grinder, ball mill, blender, and so on. Nanofibrillated cellulose produced using supermasscolloider is very common among many research groups. This process involves passing the cellulose fibers through a grinding machine with certain number of passes. The gap between grinding disks can be adjusted to vary the shear force used to break down cellulose into nanosized fibers. According to Wang et al. 2013 [16], nanofibrillated cellulose can be prepared from waste paper. To produce nanofibrillated cellulose, 1% of waste pulp was passed through the supermasscolloider (MKCA6-2) 20 times at 1600 rpm. According to Wang et al. [16], the cellulose nanofiber had a width of 30-100 nm. However, there was not a clear indication for the length of cellulose nanofiber in this research. Hai 2017 [20] reported nanofibrillated cellulose produced by supermasscolloider. In addition, he also proposed methods to define the length of single nanofibrillated cellulose. In his paper, he used FE-SEM and ImageJ to define the length of the single nanofibrillated cellulose. To do so, FE-SEM images were first taken with the full length of the nanofibrillated cellulose fiber, then the ImageJ was used to determine the length of nanofibrillated cellulose. However, this process is time-consuming, and a lot of samples are needed to be prepared and keen separation of cellulose nanofibrillated is needed to define the fiber length.

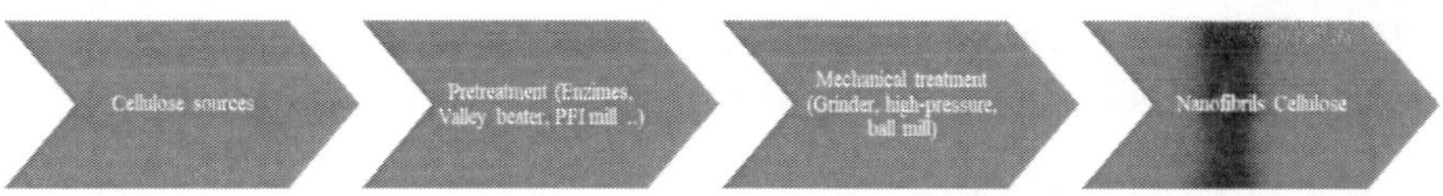

Figure 3. Illustration nanofibrillated cellulose isolation process.

Morphology by (TEM, SEM, AFM)

To evaluate the morphology and dimensions (width and length) of cellulose nanofiber, different methods such as atomic force microscopy (AFM), field emission scanning electron microscopy (FE-SEM), transmission electron microscopy (TEM), dynamic light scattering (DLS) and may others are used. There are different ways to prepare a sample for morphology determination. For AFM, samples can be dropped on a silicon wafer, or glass slide. The samples are then dried for morphology determination. Many research groups used spraying and drying the samples to determine the cellulose nanofiber morphology. The spraying and drying of the sample is usually intended to give uniformly dispersed and less dense samples in order to get individual fiber compared to dropped casting. For TEM, samples were dropped onto a coated grid and dried. TEM samples can be coated or not for morphology study. SEM and FE-SEM can also be used to determine the morphology of cellulose nanofiber. Also, spray or drop casting methods can be used to prepare the sample. For SEM and FE-SEM, it is always required that coatings such as carbon, gold or platinum be used. Besides these common processes to determine cellulose nanofiber morphology, DLS and nanoparticles analyzers were also used by different research groups.

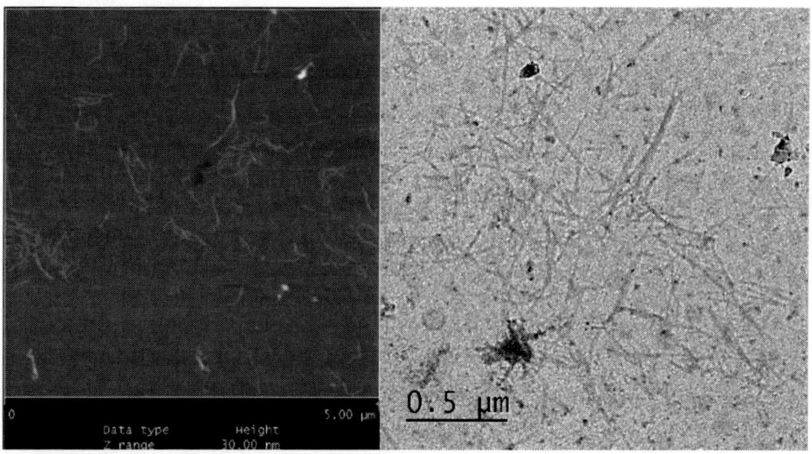

Figure 4. Cellulose nanofiber by AFM, and TEM.

Thermal Degradation Properties (TGA, TG-DTA)

Thermal stability of the material is an important factor for many applications. To sustain certain condition, higher thermal stability of a material is often preferred to thermally stable material. Therefore, cellulose and nanocellulose fiber are considered for their excellent thermal stability properties. In general the thermal decomposition temperature of cellulose is around 330-360°C. In a report, Hai et al. 2015 [8] mentioned the decomposition temperature of different cellulose source around 360°C. However, due to sulfate groups on NCC, the thermal stability of nanocrystals cellulose is usually lower. According to Kumar et al. 2014 [12], the thermal degradation of CNC was lower and in the boarder ranges also due to sulfate groups. In his study that the thermal degradation of cellulose nanocrystal was observed to begin from 220°C and up to 470°C. Nanocrystal cellulose by Mohamed et al. 2015 [18] started to degrade from around 180°C. According to his report [8, 60] the degradation of different cellulose nanocrystal starts from 185 to 230°C. To retain the thermal stability of cellulose nanocrystal, several methods were used such as using NaOH to replace the sulfate groups from cellulose. Another method to retain higher thermal stability is by inducing chemical swelling as Haafiz et al. 2014 [61], mentioned that the thermal stability of CNC is better than the thermal stability of MCC starting materials. Therefore, it is possible to achieve higher thermal stability of CNC using the above different treatments. According to Abraham et al. 2013 [62], the thermal stability of cellulose nanofiber extracted by steam explosion was lower than that of native cellulose. The degradation temperature of nanocellulose by the steam explosion was around 327°C. However, the thermal stability of nanofibrillated cellulose produced by mechanical methods such as grinding, and high-pressure homogenize is approximately equal to that of native cellulose sources. A claim by Hai 2017 [20], is that the degradation temperature of cellulose and nanofibrillated cellulose was around 300°C. Thus, it can be concluded that there is mechanical treatment has less effects to the thermal stability of cellulose nanofibrillated fibers.

Crystallinity Index (XRD)

The crystallinity index of cellulose and nanocrystal cellulose can be measured by XRD method. There are many different ways to measure the crystallinity index of cellulose. Reports mentioned that the CNC has higher CrI%. According to Kumar et al. 2014 [12], the CrI of cellulose, bleached and nanocrystals cellulose are 36%, 63.5% and reached 72.5% respectively. According to Morais et al. 2013 [63], the crystallinity of nanocrystal cellulose is 40% higher compared to that of raw cellulose material. Morais et al. 2013 [63] reported the crystallinity index or raw cellulose as 64% for raw cellulose and reached 90% for nanocellulose.

Isolation method may have effect on the CrI of cellulose nanocrystal. According to Haafiz et al. 2014 [61], the crystallinity of cellulose nanocrystal may be higher or lower than the initial cellulose materials due to the choice of treatment used. Haafiz et al. 2014 [61] also mentioned that the CrI% of microcrystals cellulose from 87% reached to 88% and 84% due to different methods of treatment employed. From Haafiz et al. [61], the CNC by acid treatment showed lower the crystalline index, but the swelling method yielded higher CrI%. Also from Maiti et al. 2013 [64], the CrI% of cellulose from different cellulose precursors could lead to different CrI development. According to Maiti et al. 2013 [64], it can be seen that the CrI of CNC from different cotton source gained higher CrI% index. However, the CrI of CNC from waste tissue paper had lower CrI value.

Some Potential Application of Nanocellulose Fiber

From previous research findings, it is evident that there are many potential applications of nanocellulose. Most of the application are listed herein as composites, coating materials, lubricants, color, pigment, reinforcement, drug-delivery, transparent film composites, transistors, sensors, solar cell and many others.

Transparent Cellulose Composite

Glass is a very much transparent material with low thermal expansion for the use in electronic devices. It presents a significant problem by being

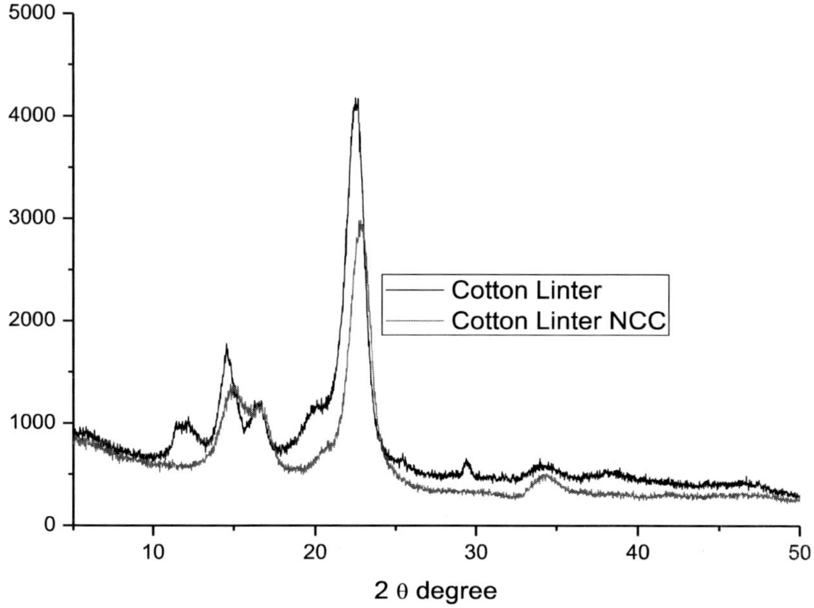

Figure 5. Crystallinity index of cellulose and nanocellulose fiber.

highly fragile. It is because of this that many researchers have embarked in investigating on a transparent material with lower thermal expansion coefficient. Paper from cellulose nanofiber can be a suitable replacement in this sector. Nogi et al. 2009 [30, 65] reported a fabrication process for transparent flexible paper from cellulose nanofiber with transparency of 80% at a wavelength of 800nm. Fukuzumi et al. 2009 [66] reported transparent film from never dried hard wood and soft wood cellulose nanofiber with optical transparency of 78% and 90% respectively in the visible light wavelength. The thermal expansion is also very low (2.7ppm K^{-1}) which is very low compared to glass. Ummartyotin et al. 2012 [67] fabricated a transparent film with a transparency of 80% in the visible light wavelength and used it as a substrate for the display of flexible organic light emitting diode (OLED).

Nanocellulose Based Hybrid Composite

The main disadvantages of biopolymer films are their poor mechanical and electrical properties for developing smart materials. Modification of polymers with inorganic materials is a new way to improve polymer properties. Hybrid composites made of nanocellulose and other materials such as graphene oxide (GO), carbon nanotube (CNT) can be made to improve important properties of the composite such as strength, electrical conductivity etc. Sadasivuni et al. 2015 [68] presented a proximity sensor by the hybridization of CNC and GO. Its working principal was demonstrated in terms of relative resistance changes in the presence of stimuli. Kafy et al. 2016 [59] demonstrated a composite prepared by simple blending of GO in CNC matrix for humidity sensor by the measurement of relative capacitance. Gao et al. 2013 [69] combined CNF and reduced graphene oxide (RGO) to make transparent conductive paper.

All-Cellulose Nanocomposite

Nowadays, people are more concerned with environmental protection. It is not only essential for human beings but also for all life on earth. Greener products; bio-degradable, and environmentally friendly products are of great interest. All-cellulose nanocomposite is a type of product that uses cellulose reinforcing in cellulose. All-cellulose composite was first named by Nishino 2004 [70]. The term was then used in many different research groups where they used cellulose reinforcing in dissolved cellulose matrix. There are different ways to prepare dissolved cellulose such as using NaOH/urea, NaOH/Thiourea, NMMO, an ionic liquid (IL), DMAc/LiCl, and several other solvents. To produce all-cellulose nanocomposite, cellulose is first dissolved in a preferred solvent, and then nanocellulose fiber is reinforced to form an all-cellulose nanocomposite. Many different approaches can be used to making all-cellulose nanocomposite. In a review by Huber 2012 [71], different types of all-cellulose composite are of high concern. Different methods to form all-cellulose composite entails using different solvents for dissolving the cellulose and nanocellulose fibers. These methods for producing all-cellulose nanocomposite include extrusion, casting, infusion, hot-pressing, and so on. From Ma et al. 2011 [72], all-cellulose composite was formed with CNC dissolved in cellulose matrix from ionic liquid solution. Also, from Ma et al. 2011 [72], by adding CNC, the crystallinity index, thermal stability, and mechanical properties of all-cellulose composite can be improved. According to Pullawan et al. 2014 [26], all-cellulose composite produced by reinforcing CNCs to cellulose matrix using NaOH/Urea and DMAc/LiCl solvent to dissolve cellulose. Pullawan et al. 2014 [26] mentioned that the tensile strength and Young's modulus of all-cellulose composite improved by adding 5 to 15% CNCs by weight. He attributed the improved mechanical properties to the high stiffness of CNC. Other studies [8-9], dissolved cellulose in DMAc/LiCl and mixed it with nanocrystals, and nanofibrils cellulose to form all-cellulose nanocomposite. The mechanical properties of the all-cellulose composite showed improvement. To produce all-cellulose nanocomposite using DMAc/LiCl, Hai et al. 2017 [9] tailored the process by exploring several ways such as using high temperature, room temperature and also solvent exchange. Cellulose matrix was dissolved and then reinforced using nanocellulose from different sources and employed different nanocellulose fabrication methods.

Acknowledgments

This work was supported by National Research Foundation of Korea (NRF-2015R1A3A2066301).

References

[1] Klemm, D., Heublein, B., Fink, H.P., Bohn A. (2005). Cellulose: Fascinating biopolymer and sustainable raw material. *Angewandte Chemie International Edition,* 44(22): 3358-3393.

[2] Rånby, B.G., Ribi, E.,1950. *Experientia* 6: 12. https://doi.org/10.1007/BF02154044.

[3] Stevanic, J., Bergstrom, E., Gatenholm, P., et al., (2012). Arabinoxylan/Nanofibrillated cellulose composite films. *Journal of Materials Science,* 47(18): 6724-6732.

[4] Nakagaito, A., Yano, H. (2004). The effect of morphological changes from pulp fiber towards nano-scale fibrillated cellulose on the mechanical properties of high-strength plant fiber based composites. *Appl. Phys. A,* 78: 547-552.

[5] Isogai, A., Kato, Y. (1998). Preparation of polyuronic acid from cellulose by TEMPO-mediated oxidation. *Cellulose, (*5): 153-164.

[6] Lee, S.Y., Chun, S.J., Kang, I.A., Park, J.Y. (2009). Preparation of cellulose nanofibrils by high-pressure homogenizer and cellulose-based composite films. *Journal of industrial and Engineering Chemist*ry, 15: 50-55.

[7] Moon, R.J., Martini, A., Nairn, J., Simonsen, J., Youngblood J. (2011). Cellulose nanomaterials review: structure, properties and nanocomposites. *Chem. Soc. Rev.,* 40: 3941-3994.

[8] Hai, L.V., Son, H.N., Seo, Y.B. (2015). Physical and bio-composite properties of nanocrystalline cellulose from wood, cotton linters, cattail, and red algae. *Cellulose,* 22: 1789-1798.

[9] Hai, L.V., Chan, H.K., Kafy, A., Zhai, L.D., Kim, J.W., Kim, J. (2017). Green all-cellulose nanocomposites made with cellulose nanofibers reinforced in dissolved cellulose matrix without heat treatment. *Cellulose*, 24 (8): 3301-3311.

[10] Iwamoto, S., Kai, W., Isogai, A., Iwata, T. (2009). Elastic modulus of single cellulose microfibrils from Tunicate measured by Atomic force microscopy. *Biomacromolecules,* 10: 2571-2576.

[11] Winuprasith, T., Suphantharika, M. (2013). Microfibrillated cellulose from mangosteen (Garciniamangostana L.) rind: preparation, characterization, and evaluation as an emulsion stabilizer. *Food hydrocolloids,* 32: 383-394.

[12] Kumar, A., Negi, Y.S., Choudhary, V., Bhardwaj, N.K. (2014). Characterization of cellulose nanocrystals produced by acid-

hydrolysis from sugarcane bagasses as agro-waste. *Journal of Materials Physics and Chemistry,* 2(1): 1-8.

[13] Okita, Y., Saito, T., Isogai, A. (2009). TEMPO-mediated oxidation of softwood thermomechanical pulp. *Holzforschung,* 63: 529-535.

[14] Lahiji, R.R., Reifenberger, R., Raman, A., Rudie, A., Moon, R.J. (2008). *Characterization of cellulose nanocrystal surfaces by SPM.* NSTI-Nanotech,www.nsti.org, ISBN 978-1-4200-8504-4, 2.

[15] Takagi, H., Nakagaito, A.N., Bistamam, M.S.A. (2013). Extraction of cellulose nanofiber from waste papers and application to reinforcement in biodegradable composites. *Journal of Reinforced Plastics &Composites,* 32(20): 1542-1546.

[16] Wang, H., Li, D., Zhang, R. (2013). Preparation of ultralong cellulose nanofibers and optically transparent nanopapers derived from waste corrugated paper pulp. *Bioresources,* 8(1): 1374-1384.

[17] Danial, W., Abdul, Majid Z., Mohd,Muhid, M., et al., (2015). The reuse of wastepaper for the extraction of cellulose nanocrystals. *Carbohydrate Polymers,* 118: 165-169.

[18] Mohamed, M.A., Salleh, W.N.W., Jaafar, J., Air, S.E.A.M., Ismail, A.F. (2015). Physicochemical properties of "green" nanocrystalline cellulose isolated from recycled newpaper. *RSC Advance,* 5: 29842.

[19] Siro, I., Plackett, D. (2010). Microfibrillated cellulose and new nanocomposite materials: a review. *Cellulose,* 17: 459-494.

[20] Hai, L.V. (2017). Properties of nanofibrillated cellulose and its length-width ratio determine by a new method. *Cellulose Chemistry and Technology,* (7-8): 549-653.

[21] Segal, L., Creely, J., Martin, A., Conrad C. (1959). An emprical method for estimating the degree of crystallinity of native cellulose using the X-ray diffractometer. *Textile Research Journal,* 29: 786-794.

[22] Favier, V., Chanzy, H., Cavaille, J.Y. (1995). Polymer nanocomposites reinforced by cellulose whiskers. *Macromolecules,* 28: 6365-6367.

[23] Azizi Samir, M.A.S., Alloin, F., Sanchez, J.Y., Dufresne, A. (2004). Cellulose nanocrystals reinforced poly (oxyethylene). *Polymer,* 45: 4149-4157.

[24] Dufresne, A. (2012). Potential of nanocellulose as a reinforcing phase for polymers, *J-For* 2(6).

[25] Li, W., Yue, J., Liu, S. (2012). Preparation of nanocrystalline cellulose via ultrasound and its reinforcement capability for poly(vinyl alcohol) composites. *Ultrasonics Sono chemistry,* 19: 479-485.

[26] Pullawan, T., Wilkinson, A.N., Zhang, L.N., Eichhorn, S.J. (2014). Deformation micromechanics of all-cellulose nanocomposites: Comparing matrix and reinforcing components. *Carbohydrate Polymers,* 100: 31-39.

[27] Luong, N.D., Pahimanolis, N., Hippi, U., Korhenen, T.J., Ruokolainen, J., Johansson, L.S., Nam, J.D., Seppala, J. (2011). Graphene/cellulose nanocomposite paper with high electrical and mechanical performances. *J. Mater. Chem.,* 21: 13991.

[28] Iwamoto, S., Nakagaito, A.N., Yano, H., Nogi, M. (2005). Optically transparent composites reinforced with plant fiber-based nanofibers. *Applied Physics A.*, 81: 1109-1112.

[29] Nogi, M., Yano, H. (2008). Transparent Nanocomposites based on cellulose produced by Bacteria offer potential innovation in the electronics Device Industry. *Advanced Materials,* 20: 1849-1852.

[30] Zhang, X., Lin, Z., Chan, B., Zhang, W., Sharma, S., Gu, W., Deng, Y. (2014). Solid-state flexible polyaniline/silver cellulose nanofibrils aerogel supercapacitors. *Journal of Power Sources,* 246: 283-289.

[31] Stephen, M., Catherine, N., Brenda, M., Andrew, K., Leslie, P., Corrine, G. (2011). Oxolane-2,5-dione modified electrospun cellulose nanofibers for heavy metals adsorption. *Journal of Hazardous Materials,* 192: 922-927.

[32] Kolakovic, R., Peltonen, L., Laukkanen, A., Hirvonen, J., Laaksonen, T. (2012). Nanofibrillar cellulose films for controlled drug delivery. *European Journal of Pharmaceutics and Biopharmaceutics,* 82: 308-315.

[33] Sun, F., Nordli, H.R., Pukstad, B., Gamstedt, E.K., Chinga-carrasco, G. (2017). Mechanical characteristics of nanocellulose-PEG bionanocomposite would dressings in wets conditions. *Journal of the Mechanical Behavior of Biomedical Materials,* 69: 377-384.

[34] Ma, H., Burger, C., Hsiao, B., Chu, B. (2014). Fabrication and characterization of cellulose nanofiber based thin-film nanofibrous composite membranes. *Journal of Membrane Science,* 454: 272-282.

[35] Kizltas, A., Kiziltas, E., Boran, S., Gardner, D. (2013). *Micro- and Nanocellulose composites for automotive applications.* 13thannual SPE automotive composites conferences and Exhibition.

[36] Alfred, C.W.L., Lam, E., Chong, J., Hrapovic, S., Luong, J.H.T. (2013). Reinforced plastic and aerogel by nanocrystalline cellulose. *J.Nanopart Res.,* 15(5): 1636.

[37] Yang, X., Shi, K., Zhitomirsky, I., Cranston, E.D. (2015). Cellulose nanocrystal aerogels as universal 3D lightweight substrates for supercapacitor materials. *Advanced Materials,* 27: 6104-6109.

[38] Lagerwall, J.P.F., Schutz, C., Salajkova, M., Noh, J.H., Park, J.H., Scalia, G., Bergstrom, L. (2014). Cellulose nanocrystal-based materials: from liquid crystal self-assembly and glass formation to multifunctional thin films. *NPG Asian Materials,* 6.

[39] Shi, Z., Phillips, G.O., Yang, G. (2013). Nanocellulose electron conductive composites. *Nanoscale,* 5: 3194-3201.

[40] Zhou, Y., Fuentes-Hernandez, C., Khan, T.M., Liu, J.C., Hsu, J., Shim, J.W., Dindar, A., Youngblood, J.P., Moon, R.J., Kippelen, B. (2013). Recyclable organic solar cells on cellulose nanocrystal substrates. *Scientific reports,* 3: 1536.

[41] Lavoine, N., Desloges, I., Dufresne, A., Bras, J. (2012). Microfibrillated cellulose – its barrier properties and applications in cellulosic materials: A review. *Carbohydrate Polymers,* 90: 735-764.

[42] Afra, E., Mohammadnejad, S., Saraeyan, A. (2016). Cellulose nanofibrils as coating material and its effects on paper properties. *Progress in Organic Coatings,* 101: 455-460.

[43] Lavoine, N., Desloges, I., Bras, J. (2014). Microfibrillated cellulose coatings as new release systems for active packaging. *Carbohydrate Polymer,* 103: 528-537.

[44] Mukherjee S.M., Woods,H.J. (1953). X-ray and electron microscope studies of the degradation of cellulose by sulphuric acid. *Biochim Biophys Acta,*10(4): 499-511.

[45] Edgar. C.D., Gray. D.G. (2001). Induce circular dichroism of chiral nematic cellulose films. *Cellulose,* 8: 5-12.

[46] Dong, X.M., Kimura, T, Revol, J.F., Gray, D.G. (1996). Effects of ionic strength on the isotropic-chiral nematic phase transition of suspensions of cellulose crystallites. *Langmuir,* 12: 2076-2082.

[47] Revol, J.F., Bradford, H, Giasson, J, Marchessault, R.H., Gray, D.G. (1992). Helicoidal self-ordering of cellulose microfibrils in aqueous suspension. *Int. J. Biol. Macromol,*14(3):, 170-172.

[48] Gaspar, D., Fernandes, S.N., de Oliveira, A.G., Fernandes, J.G., Grey, P., Pontes, R.V., Pereira, L., Martin, R., Godinho, M.H., Fortunato, E. (2014). Nanocrystalline cellulose applied simultaneously as the gate dielectric and the substrate in flexible field effect transistors. *Nanotechnology,* 25: 094008 (11pp).

[49] Santos, T.M., Souza Filho M.D.S.M., Caceres, C.A., Rosa, M.F., Moraid, J.P.S., Pinto, A.M.B., Azeredo, H.M.C. (2014). Fish gelatin film as affected by cellulose whiskers and sonication. *Food Hydrocolloids* 41: 113-118.

[50] Fahma, F., Iwamoto, S., Hori, N., Iwata, T., Takemura, A. (2010). Isolation, preparation, and characterization of nanofibers from oil palm empty-fruit-bunch (OPEFB). *Cellulose,* 17: 977-985.

[51] Gilkes, N.R., Jervis, E., Henrissat, B., Tekant, B., Miller, R.C., Warren, R.A.J., Kiburn, D.G. (1992). The adsorption of a bacterial cellulase and its two isolated domains to crystalline cellulose. *The Journal of Biological Chemistry,* 267(10): 6743-6749.

[52] Winter, H., Carclier, C., Delorme, N., Bizot, H., Quemener, B., Cathala, B. (2010). Improved colloidal stability of Bacterial cellulose nanocrystal suspensions for the elaboration of spin-coated cellulose-based model surfaces. *Biomacromolecules,*11: 3144-3151.

[53] George, J., Kumar, R., Sajeevkumar, V.A., Ramana, K.V., Rajamanickam, R., Abhishek, V., Nadanasabapathy, S., Siddaramaiah, (2014). Hybrid HPMC nanocomposites containing bacterial cellulose nanocrystals and silver nanoparticles. *Carbohydrate Polymers,* 105: 285-292.

[54] Yu, H., Qin, Z., Liang, B., Liu, N., Zhou, Z., Chen, L. (2013). Facile extraction of thermally stable cellulose nanocrystals with a high yield of 93% through hydrochloric acid hydrolysis under hydrothermal conditions. *Journal of Materials Chemistry A,* 1; 3938.

[55] Gaspar, D, Fernandes, S.N., Olivera, A.G.D., Fernandes, J.G., Grey, P., Pontes, R.V., Pereira, L., Martins, R., Godinho, M.H., Fortunato, E. (2014). Nanocrystalline cellulose applied simultaneously as the gate dielectric and the substrate in flexible field effect transistors. *Nanotechnology,* 25(9):094008. DOI: 10.1088/0957-4484/25/9/094008.

[56] Liu, Y., Wang, H., Yu, G., Li, Q., Mu, X. (2014). A novel approach for the preparation of nanocrystalline cellulose by using phosphotungstic acid. *Carborhydrate Polymers,* 110: 415-422.

[57] Saito, T., Nishiyama, Y., Putaus, J.L., Vignon, M., Isogai, A. (2006). Homogeneous suspensions of individualized microfibrils from TEMPO-catalyzed oxidation of native cellulose. *Biomacromolecules,* 7(6): 1687-1691.

[58] Saito, T., Kimura, S., Nishiyama, Y., Isogai, A. (2007). Cellulose nanofibers prepared by TEMPO-mediated oxidation of native cellulose. *Biomacromolecules,* 8: 2485-2491.

[59] Kafy et al. (2017). Cellulose long fibers fabricated from cellulose nanofibers and its strong and tough characteristics. *Scientific reports,* 7(1).

[60] Hai, L.V., Seo, Y.B. (2017). Characterization of cellulose nanocrystal obtained from electron beam treated cellulose fiber. *Nordic Pulp and Paper Research Journal,* 32(2): 170-178.

[61] Haafiz, M.K.M., Hassan, A., Zakaria, Z., Inuwa, I.M. (2014). Isolation and characterization of cellulose nanowhiskers from oil palm biomass microcrystalline cellulose. *Carbohydrate Polymers,* 103: 119-125.

[62] Ebraham, E., Deepa, B., Pothen, L.A., Cintil, J., Thomas, S., John, M.J., Anandjiwala, R., Narine, S.S. (2013). Environmental friendly method for the extraction of coir fibre and isolation of nanofibre. *Carbohydrate Polymers,* 92: 1477-1483.

[63] Morais, J.P.S., Rosa, M.D.F., Filho, M.D.S.M.S.S., Nascimento, L.D., Nascimento, D.M.D., Cassales, A.R. (2013). Extraction a characterization of nanocellulose structures from raw cotton linter. *Carbohydrae Polymers,* 91: 229-235.

[64] Maiti, S., Jayaramudu, J., Das, K., Reddy, S.M., Sadiku, R., Ray, S.S., Liu, D. (2013). Preparation and characterixation of nano-cellulose with new shape from different precursor. *Carbohydrate Polymers,* 98: 562-567.

[65] Nogi, M. (2009). Optically transparent nanofiber paper. *Advanced Materials,* 21(16): 1595-1598.

[66] Fukuzumi, H. (2008). Transparent and high gas barrier films of cellulose nanofibers prepared by TEMPO-mediated oxidation. *Biomacromolecules,* 10(1): 162-165.

[67] Ummartyotin, S. (2012). Development of transparent bacterial cellulose nanocomposite film as substrate for flexible organic light emitting diode (OLED) display. *Industrial Crops and Products,* 35(1): 92-97.

[68] Sadasivuni, K.K., Kafy, A., Zhai, L., Ko, H.U., Mun, S., Kim, J. (2015). Transparent and flexible cellulose nanocrystal/reduced graphene oxide film for proximity sensing. *Small,* 11(8); 994-1002.

[69] Gao, K., Shao, Z., Wu, X., Wang, X., Li, J., Zhang, Y., Wang, W., Wang, F. (2013). Cellulose nanofibers/reduced graphene oxide flexible transparent conductive paper. *Carbohydrate Polymers,* 91(1): 243-251.

[70] Gindl, W., Keckes, J. (2005). All-cellulose nanocomposite. *Polymer,* 46(23): 10221-10225.

[71] Huber, T., Mussig, J., Curnow, O., Pang, S., Brickerton, S., Staiger, M. (2012). A critical review of all-cellulose composites. *J Mater Sci.,* 47: 1171-1186.

[72] Ma, H., Zhou, B., Li, H.S., Li, Y.Q., Ou, S.Y. (2011). Green composite films composed of nanocrystalline cellulose and a cellulose matrix regenerated from functionalized ionic liquid solution. *Carbohydrate Polymers,* 84: 383-389.

In: Cellulose to Nanocelluloses
Editor: Tri-Dung Ngo
ISBN: 978-1-53615-057-5
© 2019 Nova Science Publishers, Inc.

Chapter 5

SIGNIFICANCE OF THE CHARACTERIZATION ACTIVITIES IN THE PRODUCTION OF CELLULOSE NANOMATERIALS

Christophe Danumah[1,*], Tri-Dung Ngo[1,2] and Behzad Ahvazi[1,3]

[1]InnoTech Alberta Inc., 250 Karl Clark Road, Edmonton, Alberta, Canada, T6N 1E4

[2]University of Alberta, Faculty of Engineering Department of Civil & Environmental Engineering, Edmonton, Alberta, Canada T6G 1H9

[3]University of Alberta, Faculty of Engineering, Department of Mechanical Engineering, Edmonton, Alberta, Canada T6G 1H9

ABSTRACT

The characterization of cellulose nanomaterials (CNMs) properties is very important for their use in different applications. This activity plays a

[*] Corresponding Author Email: christophe.danumah@innotechalberta.ca.

key role in the selection of qualified cellulose source (feedstock), monitoring, improvement and optimization of unit operation and evaluation of the quality of the produced CNMs. Profiling CNMs properties is very relevant to support applications development activities by providing high quality and consistent CNMs to recipients. To achieve this goal, state-of-art techniques are developed and utilized for profiling the internal structure and properties of the feedstocks and their derived CNMs and use the outcome to improve operation and production processes modifying equipment or unit in order to improve efficiency as well as to control the product quality. Characterization is also of a paramount importance to monitor the shelf-life and stability of CNMs as well as their precursors. Finally, it is a key path for accelerating their commercial production and application based on the fact that, information on the cellulose materials profiles allows or will allow pre-commercial and commercial producers provide their customers with high-quality, reproducible CNMs.

Keywords: cellulose nanomaterials, cellulose nanocrystals, cellulose nanofibrils, characterization, analysis, methods

INTRODUCTION

Cellulose nanomaterials (CNMs) are composed in majority by cellulose nanocrystals (CNC) and cellulose nanofibrils (CNF). Both are new classes of nanomaterials derived from forest and agriculture biomasses with a high potential economic value. To make CNMs available for applications development and later for commercial utilization, development of efficient manufacturing processes and metrology to monitor the quality of cellulose-based feedstocks and their derived CNC and CNF along with the performance of units' operation is quite essential. The measurement processes present substantial challenges since CNMs (CNC and CNF) are very new to the measurement science area.

The aim of this chapter is to provide an overview on the significance of characterization activities as an essential part of the production of cellulose nanocrystals that helped optimize CNMs production processes, yields and quality, build capability, knowledge, and expertise in the CNMs characterization area, as well as develop and co-develop methods for CNC characterization and establish protocols for quality control of CNMs.

To achieve this goal, various state-of-art characterization techniques are utilized to profile cellulose-based feedstocks and their derived CNMs (CNC and CNF) properties and improve units' operation. The profiles evaluation is of a paramount importance to monitor the shelf-life and stability of CNMs as well as their precursors. The Canadian-Standard Association (CSA) has put together a Technical Committee (TC) who has published two important standard documents such as 1) Z5100-17: Cellulose nanomaterials–Test methods for characterization [1] and 2) Z5200-17: Cellulose nanomaterials–Blank detail specification [2]. At the same time, it is also helping in developing protocols for the toxicity evaluation and the safe handling of CNMs at workplaces [3-9]. This demonstrates that the profiling of cellulose nanomaterials properties along with standardized sample preparation and analytical protocols, and safe handling are very crucial for accelerating the commercial production and application of CNMs. In addition, information on the cellulose materials profiles allows or will allow pre-commercial and commercial producers to provide their customers with high-quality and consistent CNMs.

SAMPLES PREPARATION, ANALYSIS AND CHARACTERIZATION

Samples Preparation

Samples preparation prior to analysis is very critical for material property evaluation using a specific characterization technique. This must be consistent in order to provide reproducibility and accuracy of data when comparing different samples properties using identical analysis conditions.

In other words, the sample should be representative of each and every of the feedstock prior to the analysis.

Analysis and Characterization

Analysis and characterization of cellulose-based feedstocks and CNMs (CNC and CNF) in suspensions (never dried) and in dry state (e.g., freeze-drying, spray-drying, etc.) were performed using several different state-of-art instruments in order to assess properties and access relevant information as guidance for different applications. Table 1 report major analytical techniques, physical states of materials submitted to analysis and relevant information to be extracted.

Table 1. Analytical techniques for profiling cellulose-based materials properties

Methods	Information	Physical state of Sample
Scanning electron microscopy (SEM)	Detailed nanostructural information (shape, size, aspect ratio) Dispersion state	Dry or suspended material
SEM in transmission mode (STEM)*	Detailed nanostructural information (shape, size, aspect ratio) Dispersion state Aspect ratio	Dry or suspended material
Dynamic light scattering (DLS)	Particles size and sizes distribution Aggregation state	Colloidal suspension
Zeta potential (ZP)	Nature of surface charge and density Stability Type of interactions	Colloidal suspension
X-ray diffraction (XRD) spectroscopy	Crystalline structure Crystal size Crystallinity index or degree of crystallinity	Dry material
Elemental analysis (EA) by combustion for sulfur	Total sulfur content	Dry material
Elemental analysis (EA)	Quantify surface functionalization	Dry material

Methods	Information	Physical state of Sample
Elemental analysis (ICP-OES) by acid digestion	Traces metal (Na, etc.)	Dry material
Inductively coupled plasma mass spectrometry (ICP-MS)	Quantify metal nanoparticles and CNC interaction	Colloid suspension under acid digestion
Thermogravimetric analysis (TGA/N_2)	Thermal resistance Moisture content	Dry material
Transmission electron microscopy (TEM)	Detailed nanostructural information (shape, size, aspect ratio) as well as interactions with other materials Dispersion state Aspect ratio	Dry, frozen or suspended material
Atomic force microscopy (AFM)	Structural evaluation (both lateral dimensions and particle height) Roughness, adhesion, nano-indent, surface potential in solid form	Dry material
^{13}C Solid state nuclear magnetic resonance (^{13}C NMR)	Structural evaluation & Crystallinity index or degree of crystallinity	Dry material
Fourier transform infrared spectroscopy (FTIR)	Spectroscopic evaluation (functional groups) Surface functionalization_either bulk (FTIR) or just_surface functionalization (ATR-FTIR)	Dry material (powder, film, solid)
Raman spectroscopy (RS)	Crystallinity index and surface functionalization and elastic modulus	Dry material
Rheology	Viscosity Degree of polymerization Viscosity under shear	Wet material (suspension, hydrogel or gel)
Physisorption of gas (BET model)	Specific surface area, pore size, pore volume	Dry material
X-ray photoelectron spectroscopy (XPS)	Surface functionalization	Dry material
Secondary ion mass spectroscopy (SIMS)	Ion fragments on the surfaces	Dry material
Powder wettability	Contact angle and surface free energy	Dry material
Auger electron spectroscopy (AES)	Quantitative surface analysis	Dry material
Fast protein liquid chromatography (FPLC)	Analyze or purify mixtures of CNC in different molecular weight	Colloid suspension

Table 1. (Continued)

Methods	Information	Physical state of Sample
Gel permeation chromatography (GPC)	Molecular weight from refractive index or viscosity	Colloid suspension
Energy-dispersive X-ray spectroscopy (EDS, EDX, or XEDS)	Localized elemental analysis or chemical characterization	Dry material
Conductivity meter	Conductivity measurement	Electrolytes in a solution
Differential scanning calorimetry (DSC)	Thermal stability	Dry material
Instron (Dynamic mechanical analysis orDMA)	Mechanical property	Film, fiber, solid
UV-Visible spectroscopy (UV-Vis) (Nanodrop)	Quantify surface functionalization by colorimetric analysis	Colloid suspension
Fluorometry	Fluorescence measurement	Liquid
Microsizer	Microparticles size study (e.g., spray dried CNC)	Dry material
Static light scattering (SLS)	Size study	Colloidal suspension
Confocal Microscopy (CM)	CNC and dye labeled biomolecules interaction	Film or fixed soft tissue on glass slide
Flow cytometry (FC)	The interaction of cell and functionalized CNC by counting and sorting cells	Colloidal suspension

* Kind of TEM at low kV (30kV).

Examples of Cellulose-Based Materials Properties Evaluation

This section will only involve CNC produced using concentrated sulfuric acid but information on characterization protocols for carboxylated-based CNC may be found in [1, 2].

TAPPI Standard Test Method for Pulp Chemical Composition

The TAPPI standard test method was used to elucidate the profiles between several sets of cellulose-based feedstock samples. The α-Cellulose, acetone extractives, and ash contents in the cellulose-based feedstocks were tested in accordance with TAPPI T203 cm-99, T204 om-97, and T211 om-93, respectively. Pulp viscosity was tested according to ASTM D1795-96. Degree of polymerization was calculated using the formula described by Heiningen [10].

The knowledge of the chemical composition of the cellulosic feedstock is very important for the quality of cellulose source to be used for the production of CNC. In other words, the quality of the feedstock has a direct impact on the yield and quality of the derived CNC. Table 2 below shows chemical compositions of different cellulosic feedstocks. The α-Cellulose content is one of the important parameters that can affect the production yield of CNC. For example, a feedstock with higher α-Cellulose content (e.g., 96.0% for dissolving pulp (DP)) will generate a higher CNC yield as compared to another with lower α-Cellulose content (e.g., 84.4% for Pulp 2) in identical acid hydrolysis conditions.

The level of the other components (especially the acid insoluble lignin) when higher than 4% can act as impurities and disturb the acid hydrolysis process [1]. The α-Cellulose content can be increased using appropriate pre-treatment methods but care must be taken to avoid damage and conversion of the crystalline structure of the cellulose from I (the desired one) to II.

Inductively Coupled Plasma Optical Emission Spectrometry (ICP-OES) for Sodium Content Analysis by Digestion

ICP-OES (also referred to as Inductively Coupled Plasma Atomic Emission Spectroscopy or ICP-AES), is an analytical technique used for the detection of trace metals. This technique was used to quantify traces of Na in cellulose-based feedstocks. Total Na content in solution was analyzed using an iCap 6000 series ICP-OES. Calibrations and instrument quality control checks are performed using (separate) NIST-traceable standard solutions purchased from SCP Scientific. QC checks and analysis repeats are performed on every 10 samples. An internal standard containing Yttrium (Y) is used to correct for matrix effects.

The digestion of untreated and treated pulps was performed following EPA 3051a (using nitric acid only). 5mL of trace-metal-grade 10% HNO_3 is added to a small amount of weighed sample in a Teflon reaction vessel, then heated in a MARS 5 Xpress Microwave Assisted Reaction System. After cooling, 20mL of DI: RO water (18M-ohm resistance) is added to each sample, for a final volume of 25mL. This aqueous digest is then analyzed by ICP-OES. For Na analysis, an in-house lab reference material is also periodically digested and analyzed to verify digestion performance.

This analytical technique also referred as ICP-AES. The technique is used to reveal the level of trace metal (e.g., Na) that can act as impurities (e.g., Na level) and significantly impact the quality of the cellulose-based feedstocks and in return have a negative impact on the derived CNMs quality, production yields and consistency. For example, two cellulose sources with similar α-cellulose contents but different level of Na, one higher and the other lower will generate CNC with yields higher and lower, respectively, in identical acid hydrolysis conditions. Table 3 shows Na contents in different cellulose-based feedstocks. When this content is higher as for Pulp A, it affects the acid hydrolysis process by lowering the production yields of CNMs acting as impurities. It can then be used to monitor and select the appropriate cellulose source.

Table 2. Chemical composition of different pulps

Sample	Solid content (%)	α-Cellulose content (%)	Acid insoluble lignin (%)	Ash content (%)	Acetone extractive (%)
DP	97.41	96.00	3.58	0.13	0.051
Pulp 1	96.87	92.52	1.83	0.56	0.356
Pulp 2	97.30	84.37	3.87	0.50	0.069
MCC*	97.16	63.13	11.62	0.06	0.013

Dissolving pulp (Neucel, BC, Canada); MCC: Microcrystalline cellulose (Sigma Chem, China); Pulps 1 and 2 cannot be disclosed for confidentiality issue; * MCC is neither wood nor pulp, the TAPPI method maybe not fitted for MCC sample.

Table 3. Level of Na in different cellulose-based feedstocks

Pulp ID	Na content (µg/g)[1]
Pulp 1	81.2
Pulp 2	56.1
Pulp A	296.0

[1]Elemental analysis by Inductively Coupled Plasma Optical Emission Spectrometry (ICP-OES).

Dynamic Light Scattering (DLS)

Particle size measurements were performed on never dried and spray-dried derived CNC suspensions in a temperature-regulated cell at a temperature of 25.0 ± 0.1°C using a Malvern Zetasizer Nano-S instrument working at a 173° scattering angle. This optimum angle was selected by the instrument maker, taking into account the inverse relation between particle sizes and scattering angle. The scattering power of nano-sized particles at a large scattering angle contains enough information for a general interpretation procedure. This instrument is equipped with a 4.0mW He–Ne laser (k = 633nm) and an Avalanche photodiode detector. Size distributions were calculated using an inverse Laplace transform algorithm, and the hydrodynamic radii were calculated using the Stokes-Einstein equation. DLS measures translational diffusion coefficients and then the particles size values quoted as average hydrodynamic diameters correspond to average hydrodynamic diameters of equivalent spheres and as such, they do not represent actual physical dimensions of the anisometric rod-like CNCs. However, they can be used for comparison purposes.

The concentration of the CNC suspensions was 0.1wt%. The resulting suspensions sonicated for 3min with a Branson 5510 sonicator without filtration prior to analysis. When needed, sonication was used to break down the size of aggregates/agglomerates.

The average hydrodynamic diameter of the CNCs also known as average particles size and its distribution are important for quality control, production process repeatability and grade comparison. This information is also crucial for detecting presence of CNC agglomerates and aggregates. This analysis can also be used to monitor the size reduction (breakage of agglomerates/aggregates) after submitting the CNC in suspension to high shearing forces (homogenization) or cavitation forces (sonication) as shown by CNC 6 (162.4nm before homogenisation) and CNC 7 (129.9nm after homogenisation). Both samples are from the same batch. For the latter, the average hydynamic diameter will always be lower as compared to before applying size reduction techniques. As shown in Table 4, CNC 1 to CNC 7 are samples prepared using different acid hydrolysis conditions. Emerging average hydrodynamic diameters data show that the particles size is conditions/post-processing dependent. Therefore, this analysis can be used as quality control.

Zeta Potential (ZP) Measurements

ZP is used to quantify the magnitude of electrical charge on the surface of CNC dispersed particles; which relates to the stability of CNC colloidal suspensions. ZP surface charges were evaluated using the Malvern Zeta sizer Nano Series ZS instrument. CNC suspensions used in the ZP analysis were same as the ones for DLS. The pH of CNC suspensions was approximately 6.0. This analysis was performed on either never dried suspensions or re-dispersed suspensions (from dried CNMs).

This analytical technique is used for accessing the nature and level of the value of surface charge that control the colloidal stability, rheological properties, re-dispersability and self-assembly behavior in suspensions. It can also be used to monitor the presence of new charges on the surface of CNM after modifications.

Layer of fluid attached to a colloidal particle, such as CNCs, which is related to both the surface potential and the surface charge density and is thus a measure of colloidal stability.

Zeta potential is used for quantification of the magnitude of electrical charge on the surface of dispersed particles, which relates to the stability of colloidal suspensions. It can also be used to identify the charge as positive or negative due to charged species grafted to the CNM surface (sulfate, carboxylate groups, etc.). Zeta potential is affected by the pH, the temperature, and by the presence of electrolytes in the suspension medium.

The dividing line between a stable and unstable suspension is generally taken as +30mV or -30mV, i.e., particles with zeta potential values that are more positive than +30mV or more negative than -30mV are normally considered stable.

The magnitude of the zeta potential gives an indication of the potential stability of the colloidal system.

- From 0mV to ± 10mV = rapid coagulation or flocculation;
- From ± 10mV to ± 30mV = incipient instability;
- From ± 30mV to ± 40mV = moderate stability;
- From ± 40mV to ± 60mV = good stability; and
- More than ± 61mV = excellent stability.

The surface charges and colloidal stability for 7 CNC samples as reported in Table 4 were also evaluated. Emerging data show very stable CNC suspensions with average zeta potential values in - 49.0 and - 42.0mV range as compare to less stable CNC suspension with an average zeta potential value of - 26.0mV.

Table 4. Summary of characterization of CNC suspensions prepared using different conditions by Dynamic Light Scattering (DLS) and Zeta Potential (ZP)

CNC Sample ID	DLS		ZP (mV)	pH
	d_{av} (nm)	PDI		
CNC 1	99.8 ± 4.9	0.382 ± 0.008	- 48.8 ± 1.6	7.0
CNC 2	105.1 ± 1.9	0.345 ± 0.006	- 48.8 ± 1.2	7.0
CNC 3	127.2 ± 16.0	0.475 ± 0.029	- 45.7 ± 1.5	6.5
CNC 4	307.7 ± 0.0	0.641 ± 0.000	- 26.5 ± 0.0	6.5
CNC 5	157.7 ± 21.6	0.557 ± 0.026	- 42.1 ± 1.4	6.5
CNC 6	162.4 ± 1.0	0.594 ± 0.023	- 44.0 ± 0.7	6.5
CNC 7	129.9 ± 4.3	0.562 ± 0.000	- 43.4 ± 1.0	6.5

DLS: Dynamic Light Scattering (d_{av}: average particle size); ZP: Zeta potential (surface charge); PDI: Polydispersity Index.

Elemental Analysis (EA) of Total Sulfur Content by Combustion

This test establishes the method for determining the total amount of elemental sulfur present in samples of cellulose nanocrystals (CNC) extracted by sulfuric acid hydrolysis. The total sulfur content of a CNC sample may be used to determine the purity of the sample. The total sulfur content determination by combustion was performed using EA 1108 CHNS-O elemental analyzer to represent the cellulose surface sulfation or charge.

The total sulfur content of a CNC sample may be used to determine the purity of the sample, i.e., the presence of residual free sulfate, if it is measured before and after purification of the sample to remove free ions. The total sulfur content can be converted to the surface sulfate ester content. Assuming that all sulfur is in the form of sulfate esters located at the surface of the CNCs, this test EA (CHNS) by combustion allows the average CNC surface density, which controls its colloidal stability in aqueous suspension, to be calculated using the average particle dimensions. Table 5 gives examples of total sulfur contents of CNC materials. The total sulfur contents for all CNC materials fall in 0.5-0.7wt% range (≤ 2wt%) as expected for CNC materials with the 0.5-1.0wt% as preferred range. The maximum total of sulfur content of CNC is 2wt% and will depend on the acid hydrolysis conditions.

Table 5. Evaluation of total sulfur contents in different spray-dried CNCs

Pulp ID	Total S content (wt%)[1]
CNC 1	0.63
CNC 2	0.60
CNC 6	0.70
CNC 7	0.62

[1]Elemental analysis (CHNS-O) by combustion.

Thermogravimetric Analysis under Nitrogen (TGA/N_2)

This test method allows assessing dry materials (cellulose-based feedstocks and CNMs (CNC and CNF) thermal stability through the determination of the temperature at which the material starts to decompose, and may also provide an indication of usability for high temperature processing and end use. The thermal resistance analysis was carried out under nitrogen atmosphere using a TGA model Q500 thermogravimetric analyzer. The weights of pre-treated pulp and CNC samples were in 20-50mg range. The temperature was increased from 30°C to 600°C at a heating rate of 10°C/min. The nitrogen gas flow rate was 60mL/min. All feedstocks and CNMs were in dry state prior to thermal resistance evaluation.

This test allows assessing the thermal resistance of material through the determination of the temperature at which the materials start to decompose or react in an isothermal or dynamic system, and the extent of mass change using thermogravimetry.

The evaluation of thermal degradation profile of CNM under nitrogen is called pyrolysis. The pyrolytic properties of properties of cellulose or CNM are controlled by their chemical composition as well as the presence of specific funtional groups on their surface (e.g., sulfate groups on CNC surface after acid hydrolysis of cellulose with concentrated sulfuric acid). Pyrolysis of CMs (Cellulose-based feedstocks, CNC and CNF) proceeds through a series of complex, concurrent and consecutive reactions and provides a variety of products.

The thermal properties cellulose-based feedstocks (e.g., fully bleached pulps as well as their derived CNC samples) are evaluated using TGA under nitrogen atmosphere or air. Generally, the degradation profile of cellulose-based feedstocks (e.g., fully bleached pulps) and CNC by TGA is divided in three stages. Figure 1 shows representative TGA curves of a cellulose source and its derived CNC on heating under nitrogen gas (pyrolysis).

First Stage

Both feedstock and derived CNC samples display a small weight loss at 30-150°C temperatures range as shown in Figures 1. This loss is associated with the evaporation of moisture contained in the pulp or CNC samples. This loss is followed by a plateau until the initial degradation temperature (T_{iD}) as reported in Table 6 and shown in Figure 1 for each sample (pulp or CNC) is reached. The T_{iD} of feedstock is about 275°C (generally in 275 and 300°C temperatures range) and is higher than the T_{iD} of its corresponding CNC material because of the absence of sulfate groups on the pulps surface.

Second Stage

The degradation of cellulose usually occurs after the plateau in 250 and 400°C temperatures range [11]. The latter is affected by the level of purity of the cellulose. For instance, the presence of sulfate groups onto CNC surface will lower down their T_{iD} as compared to the ones of their pure cellulose-based precursors. In addition, the higher the density of sulfate groups on CNC surface is, the lower the T_{iD} will be. TGA curves in Figures 1 show that the TiD range of pure cellulose shifted from 250-400°C to 200-315°C due to the presence of sulfate groups onto CNC material surface. In these ranges, the primary pyrolysis of cellulose that occurs is catalyzed by sulfate groups onto CNC materials surface. The presence of sulfate groups onto CNC surface is confirmed by the negative zeta potential (ZP) values as reported in Table 4.

Third and Last Stage

Pyrolysis of hydrocarbon chains occurs along with the slow charring process of the solid residue [12, 13] and correspond to 400-600°C temperatures range for pulps and 315-600°C temperatures ranges for their derived CNC materials [14-20].

The pyrolysis profiles of CNF and Cellulose are similar for both unmodified materials.

Table 6. Summary of TGA data of a cellulose-based pulp and its derived CNC

Sample	M.C. at 150°C, %	$T_{5\%}$, °C	$T_{50\%}$, °C	T_{max}, °C	Residue at 600°C, %	T_{iD}, °C
Pulp	3.4	273.6	340.7	344.8	6.8	275.0
CNC	3.7	225.6	336.0	284.8	29.5	220.0

$T_{5\%}$: Temperature of 5% weight loss. M.C.: Moisture content at 150°C. $T_{50\%}$: Temperature of 50% weight loss. T_{max}: Temperature of maximum weight loss. T_{iD}: initial decomposition or degradation temperature. Residue: material left after pyrolysis around 600°C. Consistencies are reported in Table 2.

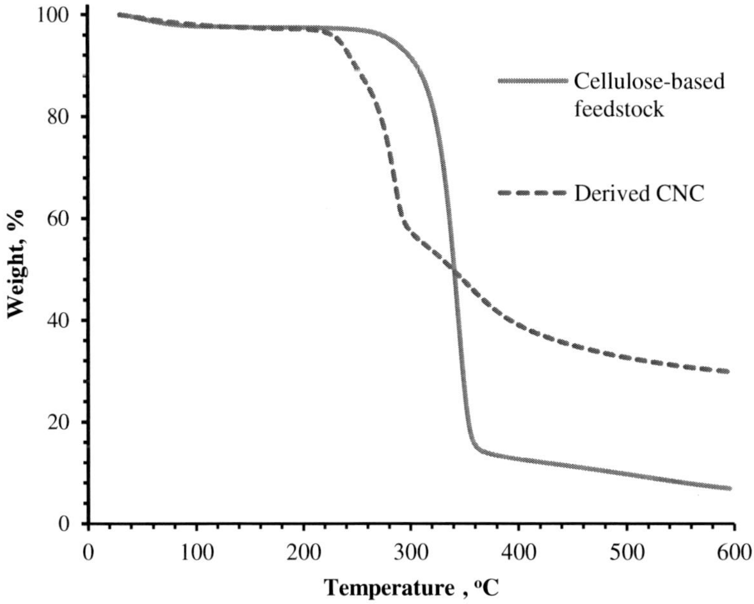

Figure 1. Representative TGA curves of a cellulose-based pulp and its derived CNC materials.

This decomposition process is very complex since it depends on quality of feedstocks (from less bleached to fully bleached, traces of impurities, etc.), density of charges (e.g., sulfate groups) onto the cellulose nanomaterials or CNMs (CNC or CNF) surface, traces of impurities in the CNMs as well as the amount of small sizes particles composing the CNMs.

X-ray Diffraction (XRD)

The XRD studies of cellulose-based feedstocks and CNMs (CNC and CNF) in dry state were recorded on a Bruker D8 Discover X-ray diffractometer equipped with CuKα source (I μS, with Montel X-ray Mirrors) at 50kV∗1000μA, 50kW over a 2θ range of 5° to 70° at room temperature, at a scan rate of 2°/min. The wave length is $\lambda = 1.5406$Å.

The simplest XRD analysis method developed by Segal et al. [21] was utilized to access the crystalline indices (CrI) of the cellulose sources CNMs samples. The CrI was calculated using the peaks height ratio based on the empirical formula [(I_{200}-I_{am})/I_{200}) x 100%], where I_{200} is the intensity of the (200) diffraction peak (at 2θ = 22.5°) and I_{am} the amorphous contribution (at 2θ = 18°). Know that the positions of the peaks at 22.5 and 18° are not absolute. This method does not account adequately for the actual position of the broad amorphous peak and uses data for only one of the several crystalline peaks. Furthermore, the peaks due to crystalline and amorphous cellulose vary in width. It is generally accepted that the empirical peak height method overestimates the degree of crystallinity.

XRD is a suitable, easy-to-use method to obtain crystalline structure and crystallinity measurements of cellulose-based materials in dry state (pulp, CNC, etc.). This test establishes the relative amount of crystalline material (mass fraction of crystallites in the bulk material) in cellulose nano/materials.

It is also a reliable and consistent method for reproducibly measuring crystallinity that can be useful for quality material control and assurance, batch control and repeatability. The degree of crystallinity can also be used to differentiate between CNFs and CNCs. It can also help understand and quantify any changes to the surface of the cellulose nanocrystal, through surface modification or other means. Because the acid hydrolysis process degrades more amorphous regions, the crystallinity index of CNC must always be higher than the one of the starting material regardless the type of method utilized to access it. Since CNF is more flexible as compared to CNC, even though its crystalline structure is of type I like CNC, its crystallinity index (degree of crystallinity) will be close to the one of the starting material. In fact, there is almost no degradation of its amorphous regions. Examples of data are shown in Table 7. CNC materials will always display cellulose I structure if a cellulose-based feedstock of cellulose I structure is used. Typical cellulose I structure is shown in Figure 2. In the contrary, the degree of crystallinity or crystallinity index (CrI) of a CNC or a CNF will depend on the one of the starting material and/or the processing conditions for their isolation. Usually, the CrI of a CNC is higher than the one of the feedstock while the CrI of CNF will be close to the one of the starting material.

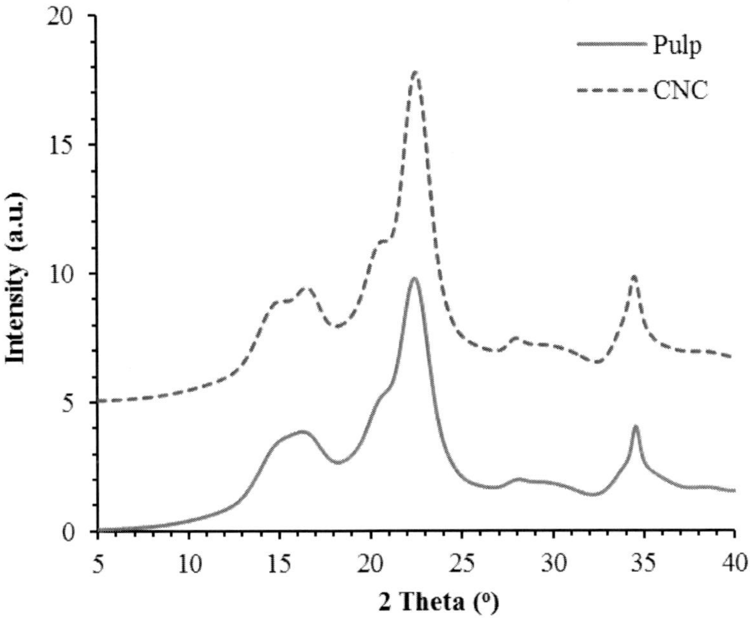

Figure 2. Representative X-ray diffraction pattern showing structural characteristics of cellulose I for cellulose-based feedstocks (e.g., fully bleached pulp) and CNMs (CNC and CNF).

Table 7. Evaluation of crystalline index of feedstocks and their derived CNC materials (they are all of cellulose I structure)

Pulp ID	Crystallinity index (%)[1]
Pulp 1	72.0
Pulp 2	69.7
Pulp 3	66.7
Pulp 6	71.1
CNC 1	76.9
CNC 2	72.7
CNC 3	72.2
CNC 6	74.7

[1] Calculated using Segal et al. [21] method.

Morphology of Cellulose-Based Feedstocks, CNC and CNF by Field-Emission Scanning Electron Microscopy (FE-SEM)

The morphologies of the CNC or CNF in water suspensions were investigated by SEM using a Hitachi model S-4800 apparatus equipped with a field emission source and operating at an accelerating voltage of 30kV in transmission mode (STEM). A drop of the appropriate concentration (0.1 and 0.0001wt% were used as concentrations) of colloidal CNC suspension (or 3min sonicated CNC suspension if required) was deposited on a carbon-coated TEM grid and allowed to evaporate and dry for at least 48h. Thereafter, the grid was stained by depositing a drop of uranyl acetate solution (2wt% in water) on the grid for 5 min. The excess solution was blotted using filter paper and the grid was allowed drying at room temperature for at least 24h prior to imaging. Copper carbon TEM grids (400mesh) and uranyl acetate (di-hydrate, 99.6%) were purchased from Electron Microscopy Sciences. The CNC suspension of 0.0001wt% concentration was prepared by diluting corresponding 0.1wt% stock suspension followed by vortex for 1-3s prior to grid preparation for evaporation.

Morphologies of dry cellulose-based feedstock, freeze-dried and spray-dried CNC were investigated by SEM while for CNC in suspensions, STEM was utilized. For the observation in dry state, the freeze-dried or spray-dried CNC materials were deposited on carbon tapes held by aluminum stubs. For each sample preparation, the material was spread on the carbon tap followed by gentle blowing with air prior to imaging.

FE-SEM technique was used to investigate the morphologies of several different cellulose-based feedstocks (pulps and powder) used for CNC production.

The outcome from FE-SEM images of cellulose-based feedstocks and CNC in Figure 3 was aligned with what is expected for each type of material:

Chopped pulp (representative for all chopped pulps) displays compacted cellulose fibers (almost no porosity) with widths in 10-40μm range with 40μm in majority for chopped pulp (~ 1mm for the 1cm x 1cm square piece).

The Whatman cotton filter paper #1 photograph shows chopped sheet with lower thickness (about 100μm) as compared to any other chopped pulp with thicker thickness (300 to 600μm). The same FE-SEM image shows less compacted cellulose fibers (higher porosity) with widths in 10-40μm range where 40μm width is major.

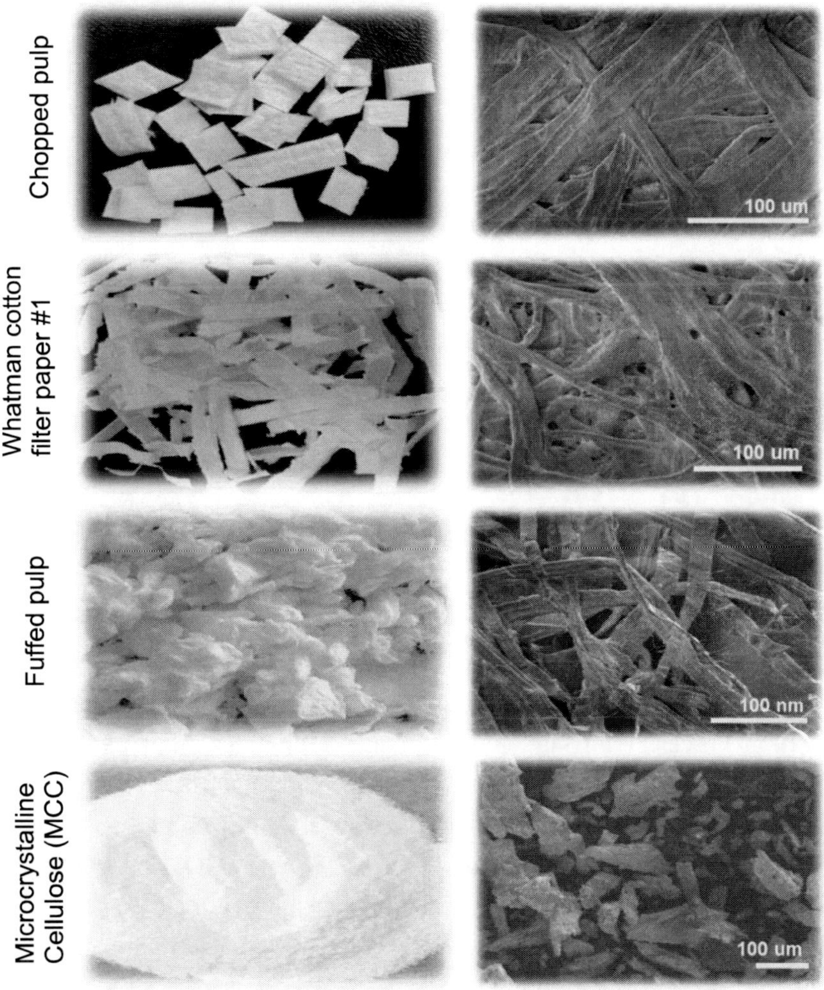

Figure 3. Representative SEM images of cellulose-based feedstocks.

For the fluffed pulp (representative all type of fluffed pulps either untreated or pre-treaed pulps), mostly individualized cellulose fibers with widths in 10-40μm range are observed on the FESEM images. Majority of the fibers have widths around 40μm.

MCC 102 powder used to previously commission the CNC pilot plant shows micro-sized (up to 20μm) particles.

Figure 4. Representative STEM images of never-dried CNC and CNF suspensions.

Overall, comparing powder, fluffed and chopped pulps, the fibers size reduction process (fluffing), more porosity (Whatman filter paper #1) or micron-sized particles for MCC (more surface area) generates more surface that is exposed to sulfuric acid. This seems to be a prerequisite for the creation of more surface area that help for a better diffusion of the acid solution along with improved interactions between the acid and the fibers surface. In return, that helps for the efficiency of the acid hydrolysis of the pulps for the CNC production.

Figure 4 shows the photographs and FE-SEM images of CNC and CNF suspensions. In general, CNC materials display rod-like structures as compared to fiber-like structures for CNF materials. The rod-like structure may sometimes look as a whisker-like structure.

Figure 5 shows photographs and FE-SEM images for representative freeze-dried and spray-dried (SD) CNC materials. FE-SEM images reveal a flake-like morphology for the freeze-dried CNC while more or less spherical particles are observed for the spray-dried CNC materials (powder). The flake-like morphology will be always observed regardless of the type of feedstock but for the spray dried CNC, the circularity of the particles will depend on the suspensions concentrations and quality (presence of impurities, aggregates/agglomerates, etc.) and the drying conditions.

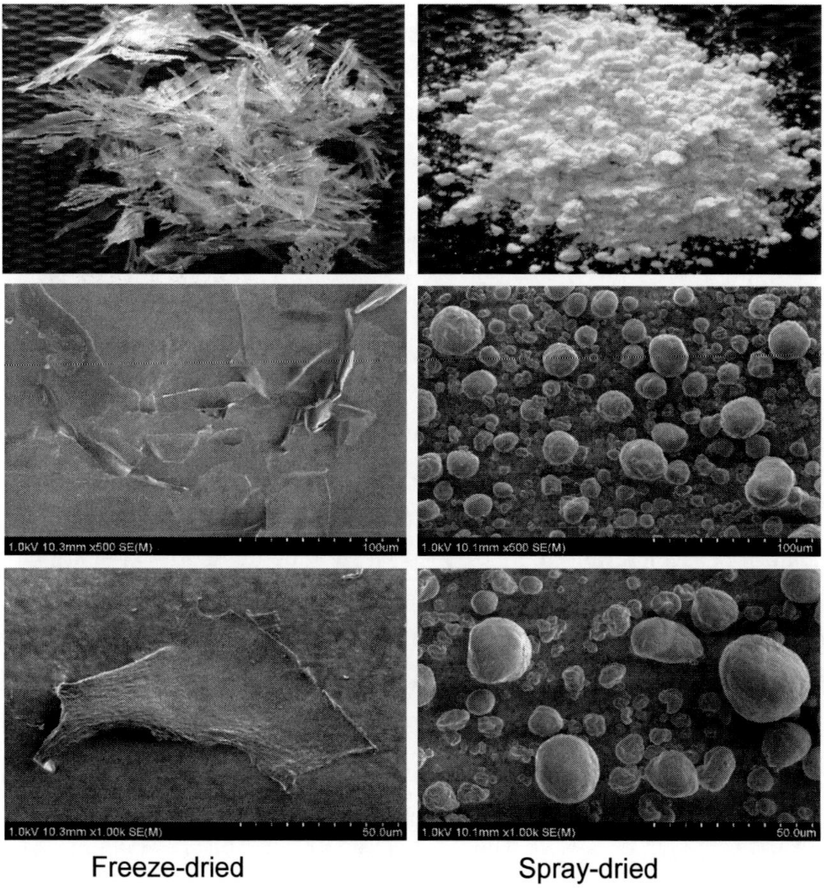

Figure 5. Representative digital pictures and SEM images of dry CNC materials.

Conclusion

The characterization activity is very crucial to identify key parameters influencing materials properties such as the crystalline structure, degree of crystallinity, thermal stability, morphology, dimensions, physical and chemical properties, CNMs production yields, and allow validate, develop and improve the physical and chemical characterization methods of CNMs.

This activity is also very relevant to propose, develop and establish standard protocols for CNMs characterization through the Canadian Standard Association Technical Committee in order to help bring CNCs to market.

The knowledge of the CNM features is very relevant to provide CNMs samples in sufficient quantity and quality to recipients (universities, government laboratories, industry, etc.) for applications development with the attempts to speed up the commercial production and utilization of CNMs.

Such an exercise is contributing to establish standard test protocols for CNMs characterization to help bring CNMs to the market. Overall, characterization methods are critical to improve quality and yields of CNMs produced and establish standard analytical protocols for quality control of CNMs.

Finally, it is also relevant for the safe handling of CNMs; which is very crucial for accelerating the commercial production and application of CNMs. Information on the cellulose materials profiles allows or will allow pre-commercial and commercial producers to provide their customers with high-quality, reproducible CNMs.

Acknowledgment

The authors would like to thank Dr. Wei-Zheng Shen for her contribution in the conception of the Table 1 entitled “Analytical techniques for profiling cellulose-based materials properties”.

REFERENCES

[1] CSA Groups. (2017). *Cellulose nanomaterials-Test methods for characterization*. Z5100-17.

[2] CSA Group. (2017). *Cellulose nanomaterials-Blank detail specification*. Z5200-17.

[3] Shatkin, J. A. (2013). *Nanotechnology-Health and Environmental Risks in Perspectives in Nanotechnology*, CRC Press, Taylor & Francis.

[4] Camarero-Espinosa, S., Endes, C., Mueller, S., Petri-Fink, A., Rothen-Rutishauser, B., Weder, C., James David Clift, M. & Johan Foster, E. (2016). Opinion - Elucidating the Potential Biological Impact of Cellulose Nanocrystals. *Fibers*, *4*, 21, 2-15.

[5] Yanamala, N., Farcas, M., Hatfield, M. K., et al., (2014). *In vivo* evaluation of the pulmonary toxicity of cellulose nanocrystals: A renewable and sustainable nanomaterial of the future. *ACS Sustainable Chem. Eng*., *2*(7), 1691-1698. DOI: 10.1021/sc500153k.

[6] Roman, M. (2015). Toxicity of cellulose nanocrystals: A review, *Industrial Biotechnology*, *11* (1), 25-33. DOI: 10.1089/ind.2014. 0024.

[7] Jeevanandam, J., Barhoum, A., Chan, Y. S., Dufresne, A. & Danquah, M. K. (2018). Review on nanoparticles and nanostructured materials: history, sources, toxicity and regulations, *Beilstein J. Nanotechnol.*, *9*, 1050-1074. DOI:10.3762/bjnano.9.98.

[8] http://work.alberta.ca/occupationalhealth-safety/295.html.

[9] Eastlake, A., Rudie, A. & Geraci, C. (2014). Nanocellulose-Evaluation of the full spectrum of workplace healt and safety. *NSTI-Nanotech*, *4*, 3, 105-108.

[10] Van Heiningen, A., Tunc, M. S., Gao, Y. & DA Silva Perez, D. (2004). Relationship between alkaline pulp yield and the mass fraction and degree of polymerization of cellulose in the pulp. *J. Pulp Paper Sci.*, *30*(8), 211-217.

[11] Randriamanantena, T., Razafindramisa, F. L., Ramanantsizehena, G., Bernes, A. & Lacabane, C. (2009). Thermal behaviour of three woods of Madagascar by thermogravimetric analysis in inert atmosphere. In:

Proceedings of the Fourth High-Energy Physics International Conference, August 21-28, 2009, Antananarivo, Madagascar.

[12] Ping-Sheng, L., Qiang, C., Xiang, L., Bo, Y., Shi-Shan, Y. W., Jian, S. & Si-Cong, L. (2009). Grafting of zwitterion from cellulose membranes via ATRP for improving blood compatibility. *Biomacromolecules*, *10*(10), 2809-2816.

[13] Jie, Y., Qunxing, X., Xuefei, Z. & Hailiang, Z. (2009). Temperature-induced chiral nematic phase changes of suspensions of poly(N,N-dimethylaminoethyl methacrylate)-grafted cellulose nanocrystals. *Cellulose*, *16*(6), 989-997.

[14] Shafizadeh, F. (1982). Introduction to pyrolysis of biomass. *Journal of Analytical and Applied Pyrolysis*, *3* (4), 283-305.

[15] Shafizadeh, F. (1984). The chemistry of pyrolysis and combustion. *Advances in Chemistry*, *207*, 489-529. DOI: 10.1021/ba-1984-0207.ch013.

[16] Van de Velde, K. & Kiekens, P. (2002). Thermal degradation of flax: The determination of kinetic parameters with thermogravimetric analysis. *J. Appl Polym Sci.*, *83*, 2634-2643.

[17] Roman, M. & Winter, W. T. (2004). Effect of sulfate groups from sulfuric acid hydrolysis on the thermal degradation behavior of bacterial cellulose. *Biomacromolecules*, *5*, 1671-1677.

[18] Wang, N., Ding, E. & Cheng, R. (2007). Thermal degradation behaviors of spherical cellulose nanocrystals with sulfate groups. *Polymer*, *48*, 3486-3493.

[19] https://web.anl.gov/PCS/acsfuel/preprint%20archive/Files/28_5_WASHINGTON%20DC_08-83_0285.pdf.

[20] George, J., Ramana, K. V., Bawa, A. S. & Siddaramaiah. (2011). Bacterial cellulose nanocrystals exhibiting high thermal stability and their polymer nanocomposites. *International Journal of Biological Macromolecules*, *48*, 50-57.

[21] Segal, L., Creely, J. J., Martin, A. E. Jr. & Conrad, C. M. (1959). An empirical method for estimating the degree of crystallinity of native cellulose using the X-ray diffractometer. *Textile Research Journal*, *29* (10), 786-794.

In: Cellulose to Nanocelluloses
Editor: Tri-Dung Ngo
ISBN: 978-1-53615-057-5
© 2019 Nova Science Publishers, Inc.

Chapter 6

NANOCELLULOSE DERIVED BIONANOCOMPOSITES

Muhammad Zubair, Rehan Ali Pradhan, Muhammad Safder, Muhammad Arshad and Aman Ullah*
Department of Agriclutural, Food and Nutritional Science, University of Alberta, Edmonton, Alberta, Canada

ABSTRACT

A Green and natural renewable materials have increasingly become a research foucs over the last two decades due to environmental issues and non-renewable petroluem resources. Cellulose nanocrystal (CNC) and cellulose nanofiber (CNF) are the materials of choice to make eco-friendly bionanocomposites with enhanced mechanical and thermal properties for various applications. These bionanocomposites possesses several attractive properties such as biodegradibility, non-toxicity and bio-based origin. This chapter highlights processing strategies and common factors affecting the CNC/CNF based bionanocomposite preparation. It also covers the recent research on cellulose-reinforced bionanocomposites, where CNF or CNC

* Corresponding Author Email: amanullah@ualberta.ca.

is used as a nanophase in synthetic or bio-based polymers or their derivatives. The mechanical and thermal proprties of the CNC and CNF derived bionanocomposites are discussed in detail. Finally, the future perspective and current challenges related to CNC and CNF based bionancomposites are outlined.

Keywords: cellulose nanocrystal, cellulose nanofiber, bionanocomposites, thermo-mechanical properties

INTRODUCTION

Presently, both natural and petroleum resources are being extensively used to produce polymers for numerous applications [1-4]. The use of these polymers in various fields have surpassed than other materials such as metals, glasses and ceramics [3, 5, 6]. Synthetic polymers as compare to natural polymers are being commonly used due to their light weight, mechanical strength, inertness and cheap production [7-9]. Though, the excessive usage of these petro-based polymers arises ecological and environmental issues [10] with their high molecular weight, chemical stability and relatively low surface area-to-volume ratio, makes them resistant to microbial degradation and cause them to persist in the environment for several hundred years [8, 11]. The conservation of petroleum resources and reduction of environmental pollution can be eased by using renewable resources which can produce green and sustainable polymers [12, 13].

Fully green polymer matrices such as protein, lipids, starch and chitosan (CS) are the alternate to petroleum based products [14-16]. Biodegradable nature of biopolymers makes them one of the favorite choices to be utilized for the preparation of environment friendly materials [17]. These bio-based materials can be easily decomposed into organic by-products (carbon dioxide and water) by the action of microorganism [14]. However, the usage of these biopolymers as materials have several drawbacks which include lower mechanical strength, thermal stability, and poor barrier properties compared to conventional materials derived petroleum resources. To overcome these issues in bio-based materials, the researchers are exploiting nanotechnology where nanofillers are used as a high performance materials in bio-based composites to impart enhanced thermal and mechanical properties, improved barrier capability, flame retardance and thermal resistivity in contrast to conventional composites [18, 19].

Bionanocomposites are a new class of nano-structured materials [20] containing the components of biological origin and particles having at least one dimension in nanometer scale (1–100 nm) [21, 22]. The existence of constituents from biological origin and nanosized material altogether provides these biomaterials with improved mechanical strength, thermal stability, and gas barrier properties, as well as an excellent biocompatibility, non-fouling properties, and biodegradability [11, 23-25].

Nanocellulose family is one of the suitable candidate to be used as a nanofiller, can be produced from cellulose [6, 26, 27]. Cellulose, the most abundant polysaccharide on earth, is abundantly available, and is ubiquitous in nature [28, 29]. Cellulose can be widely used to produce low cost, biodegradable, low density and high strength cellulose nanoparticles. Fundamentally, two types of nanoreinforcement can be obtained from cellulose: Cellulose nanofiber (CNF) (also named as nanofibrillated cellulose, microfibrillated cellulose, or cellulose nanofibrils) [30] and cellulose nanocrystal (CNC). These nano-structured materials are the favorite choices for the preparation of green bionanocomposites [15, 31] due to their high surface area, aspect ratio, and crystallinity. These characteristics of CNC and CNF allow the formation of a percolation network (in case of CNC) or an inherent entangled network (in case of CNF) morphology. These factors act as an aid to enhance mechanical, thermal, and barrier properties of polymer matrices [32, 33].

Cellulose nanomaterials (CNs) are group of non-toxic, green and sustainable bio-based materials nearing commercialization in many economic sectors, markets and products with a potential to replace petroleum based materials in many applications [26, 34]. Nevertheless, there are some issues related to CNC and CNF based materials. Particularly, nanocellulose is not compatible with most of the polymers that restrict its applications [20, 35]. Also, industrial scale production of nanocellulose derived materials is required for their utilization in various applications.

The chapter starts with brief introduction of different methods for the bionanocomposites preparation and various factors affecting their properties. Furthermore, chapter describes recent research and progress on the development of CNF or CNC based bionanocomposites. Concurrently, the role of CNF and CNC in these bionanocomposites is discussed in regard to their mechanical and thermal properties.

PROCESSING STRATEGIES FOR CELLULOSE DERIVED BIONANOCOMPOSITES

The most common composite processing methods are shown in flow chart (Figure 1) which include solvent and melt treatment [36]. In both these processes, the forces and factors are entirely different which basically affect the organization of nano particles during modification of the matrix. The solvent process is further consists of three different types of methods including Casting/evaporation, template and electrospining process.

Solvent Casting/Evaporation

Solvent casting/evaporation is the common reported method for the preparation of nanocomposite films. Normally the concentration of CNC used is in the range of 0.05 to 5%. The dispersion of CNC in polymer matrix and in solvent medium is very important for achieving desired properties. Generally, CNC are found as water suspension which creates swelling of the cellulose by saving energy, time and chemicals [37]. While drying of suspension causes irreversible aggregation of CNC which will ultimately affect the nano structure of cellulose. Now a days, the solvent casting is not only limited to water but other organic solvents are also being used i.e., pyridine, toluene, chloroform and dimethylformaide (DMF) [38]. For instance, Asad et al. prepared nanocellulose derived composite films from oil palm pulp by solvent casting method [39].

Template Process

This process is a modification of solvent casting process except CNC network is made before evaporation of solvent. In case of template process the slow exchange from moisture to non-solvent forms the nano-cellulose gel. The advantage of this process is that it gives neat cellulose based films while the disadvantage of the template process is that it is lengthy method [40].

Electrospinning Method

Electrospinning method is also a type of solvent casting as this process starts with the dispersion of CNC in the required medium and then the polymer matrix is mixed with the CNC. The evaporation of the solvent take place simultaneously when the mixture passed through the collectors and the resulting composite is obtained. The advantage of this method is, one can get controlled thickness of the composite films which provides high surface area to volume ratio [41]. The preparation of lignocellulosic composites is reported by this process [42].

Melt Processing

Melt processing is the most recent technique in which only temperature is required for the preparation of composites. This process is considered as eco-friendly and most acceptable process for the industry because of no solvent usage. This process is fast as it takes minutes compared to solvent casting which required several hours [43]. Le et al., prepared cellulose nanocomposites by melt processing using cellulose nanofibril and polycaprolactone [43].

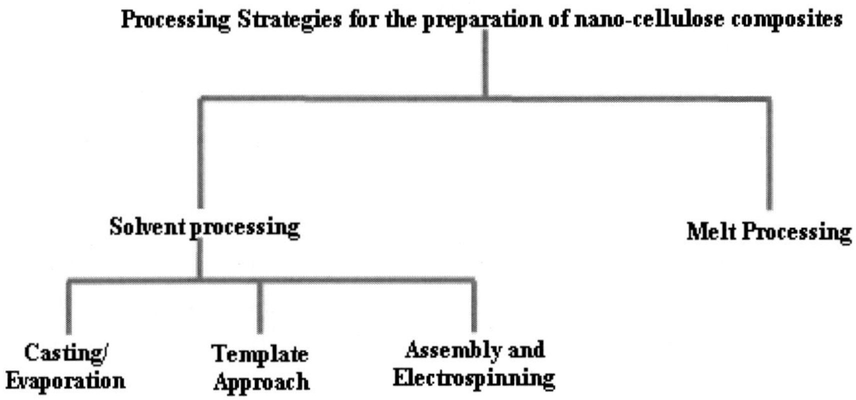

Figure 1. Flow chart showing processing methods for the preparation of nano-cellulose composites.

CHALLENGES ASSOCIATED WITH THE USE OF CELLULOSE NANOCRYSTALS (CNC)

Although, on one hand the largest surface area of CNC is advantageous, while on the other hand it is a big challenge because it results in the aggregation of CNC [44]. CNC molecules get closer to each other because of hydrogen bonding in between hydroxyl functional groups present on their surface which eventually leads to their aggregation [45]. To avoid this issue the most specific controlled processes should be followed. The distance between CNC layers is in the range of 1.2 to 1.6 nm, this small gap creates huge challenge that these layers maintain distance and do not aggregate [46]. Secondly, the polar surface of cellulose creates another challenge when these are mixed in non-polar solvents, the attraction between hydroxyl groups become stronger which results n the aggregation of CNC molecules [45]. To overcome this issue different physico-chemical processes are needed such as surface modification of CNC which results in the reduction of steric effect among them [47]. The other method to minimize aggregation is the grafting of non-polar groups on the surface of CNC or by adsorbing surfactants [48]. All of the above mentioned methods decrease the internal energy of CNC which eventually decrease the aggregation.

Aggregation of CNC also has effect the end use of composites which create void/holes because the CNC are not homogenously dispersed into the matrix. Aggregation also affects the barrier properties of composites because the diffusion of small molecules becomes easier [45]. The other drawback of aggregation include reduction in the mechanical and thermal properties because higher temperature is required for the thermal degradation of aggregates which potentially leads to the decomposition of the matrix [49].

CELLULOSE BASED BIONANOCOMPOSITES

Cellulose Nanofibers Derived Bionanocomposites

Cellulose nanofiber (CNF) is a sustainable, renewable, green, biodegradable and abundantly available nanofiller to make environment friendly bionanocomposites. CNF as a nanofiller, has been most commonly used into synthetic polymer, especially PLA [36, 50-59] to improve its physical properties. Natural polymers such as starch [60-65] and chitosan [66-69] based bionanocomposites have also been developed by incorporating cellulose nanofibers. There are few other studies existing on polyurethane/CNFs [70, 71] and whey protein isolates WPI/CNFs [72] based bionanocomposites. In general, mechanical and thermal properties of the bionanocomposites can be determined by the following factors: polymer matrix crystallinity, adhesion and compatibility between the nanofiller and polymer and nanofiller efficiency of stress transfer, its volume fraction, aspect ratio and orientation [73]. In this section, we discuss CNF reinforced bionanocomposites with emphasis on the mechanical and thermal properties.

PLA/CNF Based Bionanocomposites

Polyesters are produced via polycondensation reaction of polyhydric alcohols and polybasic acid. They are extensively used in various industrial applications and human daily life. They have sustainable, have good performance with low harmful effects on the environment [74]. However, processing methods for polyesters preparation are expensive and their resistance to temperature is low, gas transportation is high, easily breakable and low degradation rate which restrict their usage.

Polyester such as PLA, a green material, is one of viable choices to make composite materials as it can be produced from the agricultural biomass [75]. There are some drawbacks related to these polymers i.e., weak physical properties and rate of crystallization is slow. As a result, low degree of crystallinity is obtained, particularly during introduction of D–L lactide into the polymer matrix chain. In addition to that, PLA brittleness and low heat resistance make it difficult to replace commonly used synthetic polymers. Therefore, a number of studies have been done so far to enhance PLA properties by adding different nanofillers including CNFs.

PLA nanocomposites/CNCs bionanocomposites [54] were prepared by using extrusion and injection method. The thermal, mechanical and rheological properties were studied with the help of thermo-gravimetric analysis (TGA) and universal testing machine. Mechanical strength of the PLA was improved with the addition of 20% CNF from 62 MPa to 65 MPa. Further, addition of 1.3% weight of bioadimide also enhanced the PLA strength slightly but tensile modulus was increased up to 26%. The improvement was due to strong interactions between wood fibers and PLA which was facilitated by bioadimide and increased better interfacial adhesion between them.

A novel modified cellulose nanofiber/PLA bionanocomposites were prepared by solution casting process [53]. The CNF was modified using oleic acid and its effects on the thermal, morphological and mechanical properties of the bionanocomposites were studied. Modified CNF was introduced with concentration of 4 wt.%, 8 wt.% and 12 wt.%. Thermal behavior was evaluated with differential scanning calorimetry (DSC), which showed higher melting temperatures of the bionanocomposites as compare to films obtained by using PLA only. Modified CNF/PLA bionanocomposite with 12 wt.% concentration demonstrated highest tensile strength and young's modulus i.e., 2.5 and 2 times higher than pure PLA films. The morphological images were taken by field emission SEM fracture, showed CNF/PLA interface adhesion as compare to unmodified CNF. The FTIR measurements indicated that the modification results in the reduction of surface polarity as compare to unmodified CNF. Thus, the compatibility between PLA and CNF was improved while mechanical and thermal properties were increased.

Figure 2 (a, b, c and d) shows the stress curves, Young's modulus, ultimate strength and work of fracture respectively of the nanocomposites obtained by addition of nanocellulose fibril in different wt.%. The nanocomposite containing only 0.1 wt.% cellulose nanofibrils showed better mechanical properties than those of the neat PLLA film, and the properties were continuously improved up to 1.0 wt.% of cellulose nanofibril contents loading. The Young's modulus, ultimate tensile strength, and work of fracture of the nanocomposite containing 1.0 wt.% cellulose nanofibrils was increased by 40%, 26%, and 52% respectively as compared with the neat PLLA film [76].

In another study, cellulose nanofibers were modified into PEG-grafted cellulose nanofibers before introducing into the poly (L-lactide) PLLA matrix [56]. Cellulose nanofibril surface modification was done using amine-terminated poly(ethylene glycol). Bionanocomposite films were synthesized with chloroform using solvent casting method. Differential scanning calorimetry and polarized optical microscopy was used to study crystallization kinetics related to isotherm and non-isotherm. The significant improvement was observed in crystallization rate of PLLA after PEG addition. This change was attributed to the dense immobilization of PEG chains on the cellulose nanofibril surface. Because of the greater degree of crystallinity after isothermal crystallization, PEG-CNF/PLLA bionanocomposites had better heat resistance property than pure PLLA films.

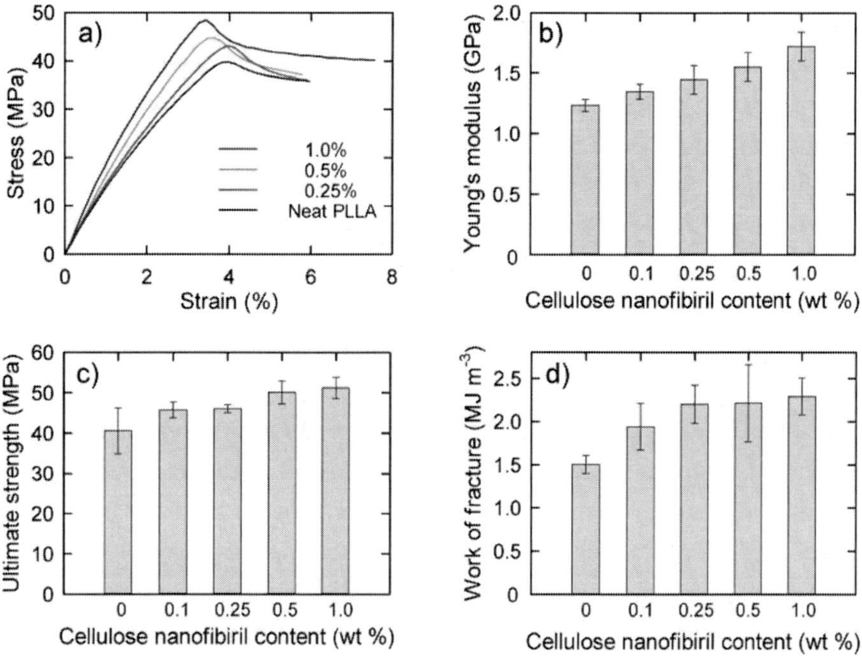

Figure 2. Stress (a), Young's modulus (b) Ultimate strength (c) and Work of fracture (d) of the neat PLA film and PLA–MCNF nanocomposites. Reproduced after permission from ACS [76].

Abdulkhani et al. [52] also did surface modification of CNFs prior to introduce into the PLA matrix. Solvent casting was used to prepare CNF/PLA green bionanocomposites. Modified CNF with 1, 3, 5 wt.% was loaded and their effect on the mechanical and thermal properties of the bionanocomposites was investigated. Scanning electron micrographs showed uniform distribution of CNF when used with 1 and 3 wt.% concentration, while aggregation was observed with 5 wt.% into the PLA matrix. The addition of 1% CNF did not have any noticeable effect on tensile strength (TS) and elastic modulus (EM) of the bionanocomposites while elongation percentage (E) was enhanced significantly, that is more than 60% to the pure PLA film. Addition of 5 wt.% CNF into PLA results in increase of modulus from 1.17 GPa to 2.12 GPa and tensile strength (100%). A significant improvement in the elongation at break with 3 wt.% CNF and an abrupt reduction at 5 wt.% was observed. While, glass transition and melting temperatures of the CNF/PLA bionanocomposites were increased slightly by incorporating CNF. However, the films exhibited almost similar thermal behavior like pure PLA film.

A study was carried out to see micro and nano size effects of the cellulose nanofibers on the crystallization of poly(lactic acid) [58]. Different size cellulose nanofibers with varying in width on nanoscale (100 nm - 3 μm) were used to reinforce PLA. The crystallization behavior of PLA with different size of CNF were measured with differential scanning calorimetry and polarized optical microscopy. The study found that the network formation of nanofiber has main role in PLA crystallization. Due to self-assembly formation of small width cellulose nanofibers, PLA crystallization was not improved. While PLA crystallization with cellulose fiber of 60 nm width was higher than others. Furthermore, the network formation of nanofibers may also be an important factor in PLA crystallites.

In another study, electrospun cellulose nanofibers of standard diameter (2 and 20 wt.%) were used to prepare PLA based bionanocomposites [50]. The properties of these prepared composites were compared with PLA/Cellulose nanofibers (2 wt.% CNF) based bionanocomposites. Both cellulose fibers with standard diameters (CSFs) and cellulose nanofibers (CNF) were synthesized with NMMO (N-methylmorpholine-N-oxide) method. The results represents that the storage modulus of bionanocomposites with 2 wt.% CNFs was increased from 14% to 25% (between temperature range of 20°C to 80°C) and yield strength improved 50% higher at 45°C as compare to pure PLA based composites. With same weight percent of CSFs, no major change was observed in the storage modulus. Further, the thermal properties of PLA in the composites remained unaffected by the presence of CSFs or CNFs.

Cellulose nanofiber (CNF) /polylactic acid (PLA) bionanocomposites [36] were developed using twin screw extrusion method and studied by Jonoobi et al. The preparation involved premixing of a higher concentration of CNFs in master batch with PLA and a diluted final concentration of 1, 3, 5 wt.% in the extrusion process. The effect of different concentration was investigated on morphology, mechanical and dynamic mechanical properties (DMA) of the bionanocomposites. Significant changes were observed with the 5 wt.% CNF concentration. The tensile strength and modulus increased tremendously i.e., from 2.9 - 3.6 GPa and 58 - 71 MPa respectively. The DMA analysis showed that the storage modulus for all nanocomposites was improved as compare to PLA and was significant at a temperature region of 70°C. Also, a shift in the tan delta peak was observed towards higher temperatures and PLA delta peak was shifted from 70°C to 76°C with 5 wt.% CNF based PLA bionanocomposites. After two years later, they studied the effect of chemically modified acetylated cellulose nanofibers [57] with PLA to prepare bionanocomposites. Universal testing machine and dynamic mechanical analyzer was used to determine mechanical and viscoelastic properties for acetylated and non-acetylated cellulose fiber derived nanocomposites. While TGA and SEM were used to study microscopic surface fracture and thermal stability. The study concluded that the nanofibers introduction significantly improved the thermal stability and storage modulus (28 times) of the PLA bionanocomposites. However, crystallinity, thermal and mechanical properties of the acetylated CNFs based bionanocomposites were not improved significantly as compare to non-acetylated CNFs. The slight improvement was observed in the storage modulus of the acetylated bionanocomposites as compare to non-acetylated.

Cellulose nanofibers with diameters ranging from 11 nm to 44 nm were used to develop PLA bionanocomposites [51]. CNFs were treated with 3-aminopropyltriethoxysilane before adding into the PLA matrix. The thermal properties of both treated and untreated nanofibers were studied with multiple session of differential scanning calorimetry (DSC). While morphological studies of bionanocomposites was done by AFM (atomic force microscopy) technique and Peak Force QNM (Quantitative Mechanical Property Mapping at the Nanoscale). This gave information about the elastic modulus mapping at the nanoscale level. SEM images of PLA/CNC fibers are depicted in Figure 3 which shows successful reinforcement of CNCs in well directed orientation [77].

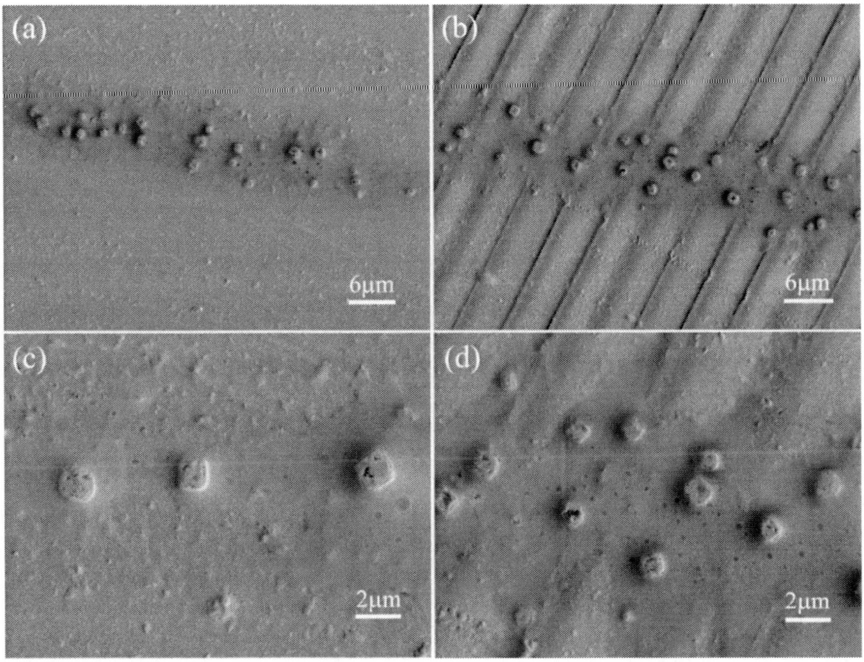

Figure 3. SEM images of PLA/CNC fibers with (a,c) 10 wt % and (b,d) 20 wt % of CNCs respectively. Reproduced with permission from ACS [77].

PLA/CNF based bionanocomposites with CNF loading of 1–2 wt.% were synthesized using solution casting method [78]. Differential scanning calorimetry (DSC) data showed that with low CNF content (1%), the CNF had no obvious effect on polymer crystallization and morphology. However, 1 wt.% of CNF addition gave excellent dispersion and a significant increase in fracture energy (200%).

Starch/CNF Based Bio Nanocomposites

CNF is also being used to alter the thermal and mechanical properties of natural polymers, such as starch. Native starch is composed of amylose and amylopectin. Starch is abundant in nature, renewable, low-cost, and environment friendly agropolymer and can be converted easily into other forms such as thermoplastic starch (TPS) to maximize starch use [26]. Due to these promising properties, starch is considered as a good substitute to synthetic polymers for various application especially packaging [79, 80]. However, the main disadvantage of starch materials is that they can absorb water easily that reduced its mechanical properties significantly.

Development of Starch/CNFs derived nanocomposites films with varying CNF weight contents from 5 to 20% by solution casting is reported [63]. The morphological, mechanical and crystallinity studies of these bionanocomposites were investigated. The results indicated that the uniform distribution of CNFs into the starch matrix was observed up to 15 wt.%, while in case of 20 wt.% loading aggregation was noticed. The CNFs addition to starch showed a great enhancement in glass transition temperature (Tg), mechanical properties including tensile strength and modulus of elasticity.

Another study [60] was reported by converting starch into thermoplastic starch first and further processed to bionanocomposites by continuous twin-screw extrusion. The study revealed the tensile testing improved by adding CNF up to 20 wt.%. Young's modulus increased linearly with the increase of CNF contents as a result highest modulus (190% increased) was observed for bionanocomposites as compared to pure TPS. However, a reduction in the elongation at break was observed by the addition of CNF. In a similar study [81], TPS/CNF nanocomposites were developed by the solution casting method. The tensile testing results showed that maximum strength was obtained at CNF content of 0.4 wt.%, afterward it started to reduce by increasing CNF contents. Agglomerations of CNF was observed at higher CNF concentration. These studies also showed that method of TPS/CNF bionanocomposite preparation influenced the dispersion of CNF in the TPS matrix and the mechanical properties.

Starch applications are limited due to its highly hydrophilic nature and lower mechanical properties as addressed by Mahyar et al. [64]. They used glycerol to plasticize the starch before adding cellulose nanofibers (0.1%, 0.2%, 0.3%, 0.4%, 0.5%, and 1%, by weight). Firstly, CNF was synthesized using short henequen fibers and, then TPS/CNF bionanocomposites were developed by solution casting method. The results indicated that 0.4 wt.% CNF/TPS films had highest tensile strength. As their tensile strength and elastic modulus went up to 80% and 170% as compare to pure TSP respectively. A reduction was seen in mechanical strength for 0.5 wt.% CNF.

Chitosan/CNF Based Bionanocomposites

Chitosan (CS) is abundant in nature, biodegradable and sustainable material. It is a linear polysaccharide having glucosamine and aminated glucosamine derived from chitin by N-deacetylation [82]. Similar to starch and cellulose, chitosan has also been extensively studied for bionanocomposite formation because of to its excellent film-forming capacity, biocompatibility, biodegradability and antimicrobial [83, 84]. Nevertheless, pure chitosan based materials have some issues such as poor chemical stability and low mechanical properties [85, 86].

Hassan et al. prepared chitosan cellulose nanocomposites and studied the effect of pretreatment of bagasse before isolation of nanofibers (cellulose) used in this study. Nanofibers (2.5 to 20 wt.%) were used with chitosan polymer to produce composite material using solvent casting technique. Alkali or xylanase treated bagasse derived nanofibers based chitosan composites demonstrated higher wet and dry tensile strength compared to those obtained by using untreated bagasse nanofibers. Altogether, improved glass transition was observed for bagasse derived cellulose based chitosan nanocomposites [66]. Hassan et al. also investigated mechanical properties of chitosan nanocomposites prepared by loading different concentrations of nanofibers (2.5 to 20 wt.%) isolated from rice straw pulp via solution casting. Increase in tensile strength and higher value of glass transition temperature was observed by these nanocomposites after addition of rice straw nanofibers to chitosan matrix. The exact role of these nanofibers is still required further studies [68]. A slight reduction in the mechanical properties of self-bonded composite films was reported by Robles et al. after addition of chitin nanocrystals in the prepared composites as unaltered young modulus and affected strain at fracture was observed [69].

In another study, adipic acid was used to dissolve and cross link the chitosan before the addition of CNF into the chitosan. The chitosan/CNF derived bionanocomposites were prepared through casting method. Adipic acid gave better tensile strength for uncured (60%) and chemically cross-linked (113%) bionanocomposites films. Furthermore, cellulose nanofibrils (CNFs) effect on mechanical properties was examined. The chitosan films plasticized with glycerol and 5 wt.% CNF showed better properties to be used as a food packaging chitosan-based materials [67].

The thermal properties of CS nanocomposites have also been reported [66, 68]. The studies indicated that the presence of CNFs into the chitosan polymer matrix became the cause of a shift in the storage modulus and tan δ peak to a higher temperature. Also, an increase in the CNFs concentration showed an increase in the storage modulus, however no significant effect was observed on tan δ peak at the glass transition temperature.

CNF Based Bionanocomposites with Other Polymer Matrices

Few studies have also been done on the polyurethane and proteins derived CNF bionanocomposites. A novel bio-based polyurethane (PU) bionanocomposites [71] with cellulose nanofiber and cellulose nanocrystals by casting/evaporation are reported. The study found that both types of bionanocomposites showed better mechanical properties as compare to pure PU. Solubility parameters determined that interfacial interaction between CNF and PU are better as compare to CNC with PU. Cellulose nanofibrils (CNF) [70] were used to prepare polyurethane (PU) bionanocomposites and their tensile strength was determined. The results showed that CNF as a nanofillers improved the tensile strength of PU up to 10 wt.% concentration. There was an increase in the glass transition and onset temperatures of the soft segment of the bionanocomposites as compare to pure PU. While crystalline phase present in the soft domains of the bionanocomposites results in reduction of its melting temperature. The effect of humidity on the PU/CNF bionanocomposites mechanical properties were also studied. The tensile strength was determined at different elongations and found a significant reduction (from 10.64 to 2.34 MPa) at 60% relative humidity for a day with 10 wt.% CNF loading.

In another study [72], cellulose nanofibers (CNF)/whey protein isolates (WPI) bionanocomposite films were produced with glycerol as a plasticizer by casting method. The effect of CNF concentration on the mechanical properties were assessed. The mechanical strength data of films exhibited nanofibers addition caused more resistant and less flexible. CNF with 4 wt.% concentration had good dispersion of the nanofibers in the whey protein matrix.

Cellulose Nanocrystals (CNC) Based Bionanocomposites

This section reviews recent progress made on thermomechanical properties of CNC based bionanocomposites with different natural and bio-derived polymer matrices such as wheat straw hemicellulose [87], protein isolated from whey [88], pectin [89], gelatin [90], carboxymethyl cellulose [91], chitosan [92], and polyesters like polylactic acid [93]. Nanocellulose based coating [94] is also mentioned herewith. The effect of various crosslinking agents and plasticizers and the concomitant effects on the mechanical properties of composites is explained after measuring their tensile properties.

In one of the studies, Pereira et al. produced biodegradable films using hemicellulose derived from wheat straw and cellulose nanocrystals, where citric acid was added as a crosslinking agent. Tensile test was performed according to D882-12 method and were compared with the help of an established model. The minimum quantity of CNCs required to attain surface interactions at interface and strengthening reinforcing effect is termed as theoretical percolation threshold. According to equation below, the theoretical percolation threshold (ϕ_v) obtained for the CNCs was 3.5 vol.% (4.4 wt.% after considering densities) with an average aspect ratio (length to width ratio) of twenty.

$$\phi_v = 0.7W/L$$

According to the regressions, a film with 15 wt.% citric acid and 4.4 wt.% CNC predicted a 14% increase in tensile strength (11.3MPa) and elastic modulus was enhanced by 48% (826 MPa), in comparison with the control film [87].

Qazanfarzadeh et al. used solution casting method to prepare composite films from whey-isolated globular protein and nanocellulose extracted from oat husk taken in varied proportions [88]. Nanocellulose was incorporated in three different fractions (2.5%, 5% and 7.5%) to see its effect on mechanical properties which were determined according to ASTM standard method D882-91. The highest tensile strength, elongation at break and Young's modulus were achieved for 5 wt.% nanocellulose. The mechanical characteristics were reduced for higher weight fraction of nanocellulose possibly due to agglomeration of the fillers.

Chaichi et al. reported nanocellulose reinforced pectin film with three different weight fractions of nanoparticles (2 wt.%, 5 wt.% and 7 wt.%) [89]. Composite film prepared with 5 wt.% filler demonstrated increase in tensile strength (13.2 MPa) as compared to control film (7.1 MPa). While further increase in the filler ratio resulted in declining of tensile strength, attributed to the agglomeration and non-uniformity in dispersion of filler within the matrix.

Mandal et al. prepared films of carboxy methyl cellulose with different weight fractions of nanocellulose [91]. Tensile tests were performed according to ASTM D638 standard method. Nanocellulose was reinforced with maximum capacity of 70 wt.% before films became fragile. This film exhibited 10% increase in tensile strength as compared to neat carboxy methyl cellulose film. Further increase in tensile strength and elastic modulus for 60/40 and 50/50 ratio of nanocellulose/carboxymethyl cellulose was observed because of greater intermolecular interaction of nanocellulose and carboxymethyl cellulose. Elongation at break and toughness was reduced by decreasing nanocellulose content. Neat nanocellulose film exhibited 10 wt.% residue at 700°C and carboxy methyl cellulose exhibited nearly 20 wt.% residue at the same temperature. Similar trend was observed in the composites. The amorphous state of Carboxymethyl cellulose led to higher weight losses during early degradation stage [91].

Further enhancement in CNC/carboxymethyl cellulose based film was demonstrated by Chi et al. through production of ternary complex consisting of cross-linked chitosan and carboxymethyl cellulose as a continuous matrix and CNC employed as a reinforcing agent [95]. The tensile strength increased with addition of 10 wt.% CNC (60.6 MPa) as compared to the control sample (43.2 MPa). Increase in Young's modulus was also observed by 52% (4.7 GPa) that was attributed to uniform dispersion of filler in the matrix and enhanced compatibility between the matrix and the nanocellulose. The measured values confirmed the synergistic effect of ternary complex attributed to high degree of crosslinking of two continuous matrices for the nanofiller.

Herrera et al. studied the effects of plasticizer and cross-linking agent on the mechanical properties of nanocellulose based coatings applied on porous paper using dip coating method [94]. Cross linking agent (citric acid) improved moisture barrier property but showed detrimental effect on mechanical characteristics. Sorbitol was added as a plasticizer that showed 37% increase in the tensile strength. Toughness value was also doubled, and the improvement was attributed to new hydrogen bond formation and stacking interaction between plasticizer and nanocellulose [96].

Mondragon et al. prepared bionanocomposite using gelatin matrix and nanocellulose. Thermal property showed remarkable improvement with the addition of nanofiller that was confirmed by increase in onset temperature of about 8°C [90]. Glass transition temperature was also shifted to higher level. Gelatin itself has high modulus of 5 GPa and nanocomposites also demonstrated similar modulus values, while slightly reduced tensile strength was observed. This was attributed to lack of stress transfer from the matrix to the filler.

Polylactide(PLA)-based nanocomposites filled with cellulose nano-whiskers(CNWs) were prepared by melt processing method. [97] PLA-g-CNW was prepared by grafting lactide on CNW containing hydroxyl functional group at the surface as shown in Figure 4. Chemical grafting technique enhanced the compatibility between hydrophilic CNW and hydrophobic PLA and improved its thermo-mechanical characteristics.

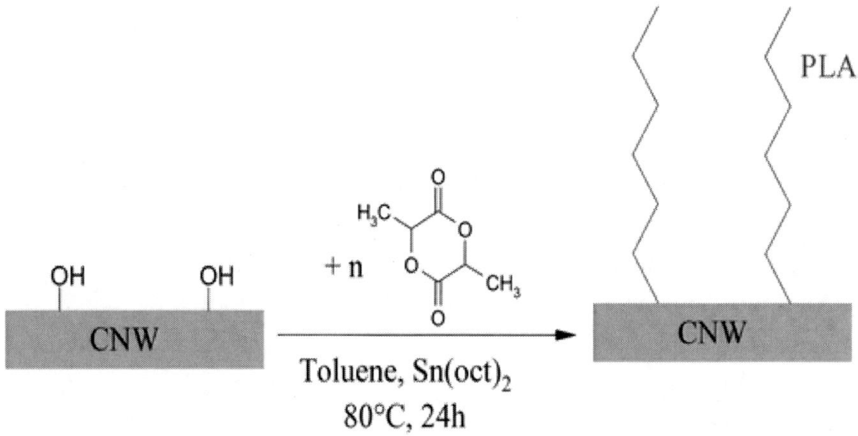

Figure 4. Grafting of PLA on the surface of cellulose nano-whiskers (CNW). Reproduced with permission from ACS [97].

Trifol et al. studied the hybrid nanocomposite of PLA with different fractions of nanocellulose and clay [93]. The crystallization behaviour of the composite was improved with 1% addition of both nanofillers and it remained unchanged with further addition. It showed remarkable improvement in their performance than neat PLA which has slow crystallization behaviour. Mechanical properties were evaluated using Instron universal testing machine are shown in Table 1. It was observed that addition of nanoclay made the film brittle and resulted in lower strain at breaking point. Composite with 5 wt.% nanocellulose and 1 wt.% clay showed significant improvement in mechanical characteristics.

Reddy et al. isolated CNC from paper-mulberry pulp and prepared bio-nanocomposite with agar [98]. Mechanical properties increased significantly even with low loading contents of CNC. Higher tensile strength of composites was (41.3 MPa with 3 wt.% CNC) obtained as compared to neat pristine agar film (33.3MPa). Elongation at break and elastic modulus were also increased and then reduction was observed on with further addition of CNC due to agglomeration of nanofiller.

Oun et al. prepared agar nanocomposite with CNC isolated from cotton linter pulp fiber [99]. Different methods were utilized for post-purification of CNC. Neutralization with NaOH resulted in enhancement of thermo-mechanical performance.

Table 1. Mechanical properties of PLA hybrid composites with different concentrations of nanocellulose crystal and clay [93]

Composition with PLA	Young's Modulus	Stress at breaking	Strain at breaking point
Neat (control sample)	2.4	55	3.9
1% CNC / 1%Clay	2.2	47	3.1
1% CNC / 3%Clay	2.3	40	2.4
1% CNC / 5%Clay	2.2	18	1.2
5% CNC / 1%Clay	2.6	49	3.0
5% CNC / 3%Clay	2.7	34	1.6
5% CNC / 5%Clay	3.0	25	1.1

CONCLUSION

Nanocellulose materials (CNC and CNF) are low-cost, green, sustainable and bio-renewable which are being used as a potential nano-reinforcement to produce bionanocomposites with excellent physical properties. CNC and CNF as a nanofillers can be introduced into synthetic polymer such as polyester, particularly PLA or in natural biopolymers such as starch, chitosan, proteins and cellulose to produce eco-friendly materials with improved mechanical and thermal properties. CNF and CNC reinforced bionanocomposites can be prepared by using various processing methods depending on the nature of matrix used. Several factors come into play during CNC and CNF based bionanocomposites formation such as pH, temperature, plasticizer and cross-linking agent and greatly influence the materials functional properties. Optimization and proper selection of process parameters are vital for the dispersion of CNC or CNF in the polymeric matrix.

Future Perspectives

Although cellulose nanocrystals and cellulose nanofibers derived bionanocomposites are in their infancy to develop renewable and sustainable materials, however, they already hold a potential to replace petroleum-based material for various applications. Recent progress is an indication towards the recognition of fully "green" nanocellulose reinforced bionano-composites. Yet, significant efforts are needed to solve two major issues. Firstly, cost-effective industrial scale preparation of nanocellulose derived materials as it is very hard for consumers to pay extra just based on the environment friendly product. Secondly, homogeneous dispersion of the nanocellulose in the matrices is a serious problem. Due to aggregation of these nanofillers, the desired properties cannot be achieved. The extensive use of CNC and CNF for multifunctional applications, collaborative research efforts from both academic and industry is highly desired to address some of the limitations related to their preparation. Nevertheless, there are challenges for the development of CNF and CNC based material. On the other hand, they have a bright prospect to broaden their applications and will be widely available to substitute petroleum-based materials in the coming future.

References

[1] Abdullah, C. K., Jawaid, M., Khalil, H. A., Zaidon, A., Hadiyane, A., 2012. Oil palm trunk polymer composite: Morphology, water absorption, and thickness swelling behaviours. *BioResources* 7, 2948-2959.

[2] Thakur, V. K., Yan, J., Lin, M. F., Zhi, C., Golberg, D., Bando, Y., Sim, R., Lee, P. S., 2012. Novel polymer nanocomposites from bioinspired green aqueous functionalization of BNNTs. *Polymer Chemistry* 3, 962-969.

[3] Thakur, V., Singha, A., 2011. Physicochemical and mechanical behavior of cellulosic pine needle-based biocomposites. *International Journal of Polymer Analysis and Characterization* 16, 390-398.

[4] Thakur, V. K., 2014. *Nanocellulose polymer nanocomposites: fundamentals and applications*, John Wiley & Sons.

[5] Silvestre, C., Duraccio, D., Cimmino, S., 2011. Food packaging based on polymer nanomaterials. *Progress in Polymer Science* 36, 1766-1782.

[6] Babu, R. P., K. O'connor, R. Seeram, Current progress on bio-based polymers and their future trends, *Progress in Biomaterials,* 2 (2013) 8.

[7] Brydson, J. A., *Plastics materials,* Elsevier, 1999.

[8] Rhim, J. W., Park, H. M., Ha, C. S., 2013. Bio-nanocomposites for food packaging applications. *Progress in Polymer Science* 38, 1629-1652.

[9] Thakur, V. K., Singha, A. S., Kaur, I., Nagarajarao, R. P., Liping, Y., 2011. Studies on analysis and characterization of phenolic composites fabricated from lignocellulosic fibres. *Polymers & Polymer Composites* 19, 505.

[10] Siracusa, V., Rocculi, P., Romani, S., Dalla Rosa, M., 2008. Biodegradable polymers for food packaging: a review. *Trends in Food Science & Technology* 19, 634-643.

[11] Rhim, J. W., Ng, P. K., 2007. Natural biopolymer-based nanocomposite films for packaging applications. *Critical Reviews in Food Science and Nutrition*. 47, 411-433.

[12] Thakur, V., Singha, A., Thakur, M., 2012. Surface modification of natural polymers to impart low water absorbency. *International Journal of Polymer Analysis and Characterization* 17, 133-143.

[13] Mohanty, A. K., Misra, M., Drzal, L., 2002. Sustainable bio-composites from renewable resources: opportunities and challenges in the green materials world. *Journal of Polymers and the Environment* 10, 19-26.

[14] Othman, S. H., 2014. Bio-nanocomposite materials for food packaging applications: types of biopolymer and nano-sized filler. *Agriculture and Agricultural Science Procedia* 2, 296-303.

[15] John, M. J., Thomas, S., 2008. Biofibres and biocomposites. *Carbohydrate Polymers* 71, 343-364.

[16] Bordes, P., Pollet, E., Avérous, L., 2009. Nano-biocomposites: biodegradable polyester/nanoclay systems. *Progress in Polymer Science* 34, 125-155.

[17] Tang, X., Kumar, P., Alavi, S., Sandeep, K., 2012. Recent advances in biopolymers and biopolymer-based nanocomposites for food packaging materials. *Critical Reviews in Food Science and Nutrition* 52, 426-442.

[18] Aranguren, M. I., Marcovich, N. E., Salgueiro, W., Somoza, A., 2013. Effect of the nano-cellulose content on the properties of reinforced polyurethanes. A study using mechanical tests and positron anihilation spectroscopy. *Polymer Testing* 32, 115-122.

[19] Thakur, V. K., Thakur, M. K., Gupta, R. K., 2014. Graft copolymers of natural fibers for green composites. *Carbohydrate Polymers* 104, 87-93.

[20] de Azeredo, H. M. C., Mattoso, L. H. C., McHugh, T. H., 2011. *Nanocomposites in food packaging-a review*. INTECH Open Access Publisher.

[21] Auad, M. L., Contos, V. S., Nutt, S., Aranguren, M. I., Marcovich, N.E., 2008. Characterization of nanocellulose-reinforced shape memory polyurethanes. *Polymer International* 57, 651-659.

[22] Ojijo, V., Ray, S. S., 2013. Processing strategies in bionano composites, *Progress in Polymer Science* 38, 1543-1589.

[23] Gironi, F., Piemonte, V., 2011. Bioplastics and petroleum-based plastics: strengths and weaknesses. *Energy Sources, Part A: Recovery, Utilization, and Environmental Effects* 33, 1949-1959.

[24] Reddy, M. M., Vivekanandhan, S., Misra, M., Bhatia, S. K., Mohanty, A. K., 2013. Biobased plastics and bionanocomposites: Current status and future opportunities. *Progress in Polymer Science* 38, 1653-1689.

[25] Pavlidou, S., Papaspyrides, C., 2008. A review on polymer–layered silicate nanocomposites. *Progress in Polymer Science* 33, 1119-1198.

[26] Endes, C., Camarero-Espinosa, S., Mueller, S., Foster, E., Petri-Fink, A., Rothen-Rutishauser, B., Weder, C., Clift, M., 2016. A critical

review of the current knowledge regarding the biological impact of nanocellulose. *Journal of Nanobiotechnology* 14, 78.

[27] Klemm, D., Heublein, B., Fink, H. P., Bohn, A., 2005. Cellulose: fascinating biopolymer and sustainable raw material. *Angewandte Chemie International Edition* 44, 3358-3393.

[28] Habibi, Y., Lucia, L. A., Rojas, O. J., 2010, Cellulose nanocrystals: chemistry, self-assembly, and applications. *Chemical Reviews* 110, 3479-3500.

[29] Mariano, M., El Kissi, N., Dufresne, A., 2014. Cellulose nanocrystals and related nanocomposites: review of some properties and challenges. *Journal of Polymer Science Part B: Polymer Physics* 52, 791-806.

[30] Podsiadlo, P., Choi, S. Y., Shim, B., Lee, J., Cuddihy, M., Kotov, N. A., 2005. Molecularly engineered nanocomposites: layer-by-layer assembly of cellulose nanocrystals. *Biomacromolecules* 6, 2914-2918.

[31] Auad, M. L., Mosiewicki, M. A., Richardson, T., Aranguren, M. I., Marcovich, N. E., 2010. Nanocomposites made from cellulose nanocrystals and tailored segmented polyurethanes. *Journal of Applied Polymer Science* 115, 1215-1225.

[32] Azizi Samir, M. A. S., Alloin, F., Dufresne, A., 2005. Review of recent research into cellulosic whiskers, their properties and their application in nanocomposite field. *Biomacromolecules* 6, 612-626.

[33] Pereda, M., Amica, G., Rácz, I., Marcovich, N. E., 2011. Structure and properties of nanocomposite films based on sodium caseinate and nanocellulose fibers. *Journal of Food Engineering* 103, 76-83.

[34] Shatkin, J. A., Kim, B., 2015. Cellulose nanomaterials: life cycle risk assessment, and environmental health and safety roadmap. *Environmental Science: Nano* 2, 477-499.

[35] De Azeredo, H. M., Nanocomposites for food packaging applications. *Food Research International* 42, 1240-1253.

[36] Jonoobi, M., Harun, J., Mathew, A. P., Oksman, K., 2010. Mechanical properties of cellulose nanofiber (CNF) reinforced polylactic acid (PLA) prepared by twin screw extrusion. *Composites Science and Technology* 70, 1742-1747.

[37] Sanchez-Garcia, M., Gimenez, E., Lagaron, J., 2008. Morphology and barrier properties of solvent cast composites of thermoplastic biopolymers and purified cellulose fibers. *Carbohydrate Polymers* 71, 235-244.

[38] Siró, I., Plackett, D., 2010. Microfibrillated cellulose and new nanocomposite materials: a review. *Cellulose* 17, 459-494.

[39] Asad, M., Saba, N., Asiri, A. M., Jawaid, M., Indarti, E., Wanrosli, W. D., 2018. Preparation and characterization of nanocomposite films from oil palm pulp nanocellulose/poly (Vinyl alcohol) by casting method. *Carbohydrate Polymers* 191, 103-111.

[40] Kargarzadeh, H., Huang, J., Lin, N., Ahmad, I., Mariano, M., Dufresne, A., Thomas, S., Gałęski, A., 2018. Recent Developments in Nanocellulose-based Biodegradable Polymers, Thermoplastic Polymers, and Porous Nanocomposites. *Progress in Polymer Science*.

[41] Cherpinski, A., Torres-Giner, S., Vartiainen, J., Peresin, M. S., Lahtinen, P., Lagaron, J. M., 2018. Improving the water resistance of nanocellulose-based films with polyhydroxyalkanoates processed by the electrospinning coating technique. *Cellulose* 25, 1291-1307.

[42] Passos de Oliveira Santos, R., Fernanda Rossi, P., Ramos, L. A., Frollini, E., 2018. Renewable Resources and a Recycled Polymer as Raw Materials: Mats from Electrospinning of Lignocellulosic Biomass and PET Solutions. *Polymers* 10, 538.

[43] Lo Re, G., Engstrom, J., Wu, Q., Malmström, E., Gedde, U. W., Olsson, R. T., Berglund, L. A., 2018. Improved cellulose nanofibril dispersion in melt-processed polycaprolactone nanocomposites by a latex-mediated interphase and wet feeding as LDPE alternative. *ACS Applied Nano Materials*.

[44] Gårdebjer, S., Bergstrand, A., Larsson, A., 2014. A mechanistic approach to explain the relation between increased dispersion of surface modified cellulose nanocrystals and final porosity in biodegradable films. *European Polymer Journal* 57, 160-168.

[45] Dufresne, A., 2013. Nanocellulose: a new ageless bionanomaterial. *Materials Today* 16, 220-227.

[46] Yang, Q., Wu, C. N., Saito, T., Isogai, A., 2014. Cellulose–clay layered nanocomposite films fabricated from aqueous cellulose/LiOH/ urea solution, *Carbohydrate Polymers* 100, 179-184.

[47] Daud, J. B., Lee, K. Y., 2017. *Surface Modification of Nanocellulose, Handbook of Nanocellulose and Cellulose Nanocomposites*, 1, 101-122.

[48] Habibi, Y., 2014. Key advances in the chemical modification of nanocelluloses. *Chemical Society Reviews* 43, 1519-1542.

[49] Isogai, A., 2013. Wood nanocelluloses: fundamentals and applications as new bio-based nanomaterials. *Journal of Wood Science* 59, 449-459.

[50] Kowalczyk, M., Piorkowska, E., Kulpinski, P., Pracella, M., 2011. Mechanical and thermal properties of PLA composites with cellulose nanofibers and standard size fibers. *Composites Part A: Applied Science and Manufacturing* 42, 1509-1514.

[51] Frone, A. N., Berlioz, S., Chailan, J. F., Panaitescu, D. M., 2013. Morphology and thermal properties of PLA–cellulose nanofibers composites. *Carbohydrate Polymers* 91, 377-384.

[52] Abdulkhani, A., Hosseinzadeh, J., Ashori, A., Dadashi, S., Takzare, Z., 2014. Preparation and characterization of modified cellulose nanofibers reinforced polylactic acid nanocomposite. *Polymer Testing* 35, 73-79.

[53] Hadi, A., Babak, G., Jalal, D., Ali, A., Asghar, K., 2015. Novel nanocomposites based on fatty acid modified cellulose nanofibers/poly (lactic acid): Morphological and physical properties. *Food Packaging and Shelf Life* 5, 21-31.

[54] Awal, A., Rana, M., Sain, M., 2015. Thermorheological and mechanical properties of cellulose reinforced PLA bio-composites. *Mechanics of Materials* 80, 87-95.

[55] Sethi, J., Farooq, M., Sain, S., Sain, M., Sirviö, J. A., Illikainen, M., Oksman, K., 2018. Water resistant nanopapers prepared by lactic acid modified cellulose nanofibers. *Cellulose* 25, 259-268.

[56] Fujisawa, S., Zhang, J., Saito, T., Iwata, T., Isogai, A., 2014. Cellulose nanofibrils as templates for the design of poly (l-lactide)-nucleating surfaces. *Polymer* 55, 2937-2942.

[57] Jonoobi, M., Mathew, A. P., Abdi, M. M., Makinejad, M. D., Oksman, K., 2012. A comparison of modified and unmodified cellulose nanofiber reinforced polylactic acid (PLA) prepared by twin screw extrusion. *Journal of Polymers and the Environment* 20, 991-997.

[58] Kose, R., Kondo, T., 2013, Size effects of cellulose nanofibers for enhancing the crystallization of poly (lactic acid). *Journal of Applied Polymer Science* 128, 1200-1205.

[59] Song, Z., Xiao, H., Zhao, Y., 2014. Hydrophobic-modified nano-cellulose fiber/PLA biodegradable composites for lowering water vapor transmission rate (WVTR) of paper. *Carbohydrate Polymers* 111, 442-448.

[60] Hietala, M., Mathew, A. P., Oksman, K., 2013. Bionanocomposites of thermoplastic starch and cellulose nanofibers manufactured using twin-screw extrusion. *European Polymer Journal* 49, 950-956.

[61] Nasri-Nasrabadi, B., Behzad, T., Bagheri, R., 2014. Preparation and characterization of cellulose nanofiber reinforced thermoplastic starch composites. *Fibers and Polymers* 15, 347-354.

[62] Yu, Z., Alsammarraie, F. K., Nayigiziki, F. X., Wang, W., Vardhanabhuti, B., Mustapha, A., Lin, M., 2017. Effect and mechanism of cellulose nanofibrils on the active functions of biopolymer-based nanocomposite films. *Food Research International* 99, 166-172.

[63] Li, M., Tian, X., Jin, R., Li, D., 2018. Preparation and characterization of nanocomposite films containing starch and cellulose nanofibers. *Industrial Crops and Products* 123, 654-660.

[64] Fazeli, M., Keley, M., Biazar, E., 2018. Preparation and characterization of starch-based composite films reinforced by cellulose nanofibers, *International journal of biological macromolecules,* 116 (2018) 272-280.

[65] Abral, H., Anugrah, A. S., Hafizulhaq, F., Handayani, D., Sugiarti, E., Muslimin, A. N., 2018. Effect of nanofibers fraction on properties of

the starch based biocomposite prepared in various ultrasonic powers, *International Journal of Biological Macromolecules* 116, 1214-1221.

[66] Hassan, M. L., Hassan, E. A., Oksman, K. N., 2011. Effect of pretreatment of bagasse fibers on the properties of chitosan/ microfibrillated cellulose nanocomposites. *Journal of Materials Science* 46, 1732-1740.

[67] Falamarzpour, P., Behzad, T., Zamani, A., 2017. Preparation of Nanocellulose Reinforced Chitosan Films, Cross-Linked by Adipic Acid, *International Journal of Molecular Sciences* 18, 396.

[68] Hassan, M. L., Fadel, S. M., El-Wakil, N. A., Oksman, K., 2012. Chitosan/rice straw nanofibers nanocomposites: Preparation, mechanical, and dynamic thermomechanical properties. *Journal of Applied Polymer Science* 125, E216-E222.

[69] Robles, E., Salaberria, A. M., Herrera, R., Fernandes, S. C., Labidi, J., 2016. Self-bonded composite films based on cellulose nanofibers and chitin nanocrystals as antifungal materials. *Carbohydrate Polymers* 144, 41-49.

[70] Sanches, A. O., Ricco, L. H. S., Malmonge, L. F., da Silva, M. J., Sakamoto, W. K., Malmonge, J. A., 2014. Influence of cellulose nanofibrils on soft and hard segments of polyurethane/cellulose nanocomposites and effect of humidity on their mechanical properties. *Polymer Testing* 40, 99-105.

[71] Benhamou, K., Kaddami, H., Magnin, A., Dufresne, A., Ahmad, A., 2015. Bio-based polyurethane reinforced with cellulose nanofibers: a comprehensive investigation on the effect of interface. *Carbohydrate Polymers* 122, 202-211.

[72] Carvalho, R. A., Santos, T. A., de Azevedo, V. M., Felix, P. H. C., Dias, M. V., Borges, S. V., 2018. Bio-nanocomposites for food packaging applications: effect of cellulose nanofibers on morphological, mechanical, optical and barrier properties. *Polymer International* 67, 386-392.

[73] Dufresne, A., Dupeyre, D., Paillet, M., 2003. Lignocellulosic flour-reinforced poly (hydroxybutyrate-co-valerate) composites. *Journal of Applied Polymer Science* 87, 1302-1315.

[74] Zia, K. M., Noreen, A., Zuber, M., Tabasum, S., Mujahid, M., 2016. Recent developments and future prospects on bio-based polyesters derived from renewable resources: A review. *International Journal of Biological Macromolecules* 82, 1028-1040.

[75] Huang, J., Chang, P. R., Chen, Y., Gao, S., Liao, J., 2017. Fully Green Cellulose Nanocomposites, *Handbook of Nanocellulose and Cellulose Nanocomposites* 1, 301-334.

[76] Fujisawa, S., Saito, T., Kimura, S., Iwata, T., Isogai, A., 2013. Surface Engineering of Ultrafine Cellulose Nanofibrils toward Polymer Nanocomposite Materials. *Biomacromolecules* 14, 1541-1546.

[77] Huan, S., Liu, G., Cheng, W., Han, G., Bai, L., 2018. Electrospun Poly(lactic acid)-Based Fibrous Nanocomposite Reinforced by Cellulose Nanocrystals: Impact of Fiber Uniaxial Alignment on Microstructure and Mechanical Properties. *Biomacromolecules* 19, 1037-1046.

[78] Okubo, K., Fujii, T., Thostenson, E. T., 2009. Multi-scale hybrid biocomposite: processing and mechanical characterization of bamboo fiber reinforced PLA with microfibrillated cellulose. *Composites Part A: Applied Science and Manufacturing* 40, 469-475.

[79] Stepto, R., 2003. The processing of starch as a thermoplastic, in: *Macromolecular Symposia*, Wiley Online Library, pp. 203-212.

[80] Angles, M. N., Dufresne, A., 2001. Plasticized starch/tunicin whiskers nanocomposite materials. 2. Mechanical behavior. *Macromolecules* 34, 2921-2931.

[81] Savadekar, N., Mhaske, S., 2012. Synthesis of nano cellulose fibers and effect on thermoplastics starch based films. *Carbohydrate Polymers* 89, 146-151.

[82] Kumar, M. N. R., 2000. A review of chitin and chitosan applications. *Reactive and Functional Polymers* 46, 1-27.

[83] Shukla, S. K., Mishra, A. K., Arotiba, O. A., Mamba, B. B., 2013. Chitosan-based nanomaterials: A state-of-the-art review. *International Journal of Biological Macromolecules* 59, 46-58.

[84] Rhim, J. W., Hong, S. I., Park, H. M., Ng, P. K., 2006. Preparation and characterization of chitosan-based nanocomposite films with

antimicrobial activity. *Journal of Agricultural and Food Chemistry* 54, 5814-5822.

[85] Coma, V., Deschamps, A., Martial-Gros, A., 2003. Bioactive packaging materials from edible chitosan polymer—antimicrobial activity assessment on dairy-related contaminants. *Journal of Food Science* 68, 2788-2792.

[86] Moreira, M. D. R., Pereda, M., Marcovich, N. E., Roura, S. I., 2011. Antimicrobial effectiveness of bioactive packaging materials from edible chitosan and casein polymers: assessment on carrot, cheese, and salami, *Journal of Food Science* 76, M54-M63.

[87] Pereira, P. H., Waldron, K. W., Wilson, D. R., Cunha, A. P., de Brito, E. S., Rodrigues, T. H., Rosa, M. F., Azeredo, H. M., 2017. Wheat straw hemicelluloses added with cellulose nanocrystals and citric acid. Effect on film physical properties. *Carbohydrate Polymers* 164, 317-324.

[88] Qazanfarzadeh, Z., Kadivar, M., 2016. Properties of whey protein isolate nanocomposite films reinforced with nanocellulose isolated from oat husk, *International Journal of Biological Macromolecules* 91, 1134-1140.

[89] Chaichi, M., Hashemi, M., Badii, F., Mohammadi, A., 2017. Preparation and characterization of a novel bionanocomposite edible film based on pectin and crystalline nanocellulose. *Carbohydrate Polymers* 157, 167-175.

[90] Mondragon, G., Peña-Rodriguez, C., González, A., Eceiza, A., Arbelaiz, A., 2015. Bionanocomposites based on gelatin matrix and nanocellulose. *European Polymer Journal* 62, 1-9.

[91] Mandal, A., Chakrabarty, D., 2014. Studies on the mechanical, thermal, morphological and barrier properties of nanocomposites based on poly (vinyl alcohol) and nanocellulose from sugarcane bagasse. *Journal of Industrial and Engineering Chemistry* 20, 462-473.

[92] Dong, F., Li, S., Jin, C., Liu, Z., Zhu, K., Zou, H., Wang, X., 2016. Effect of nanocellulose/chitosan composite coatings on cucumber

quality and shelf life. *Toxicological & Environmental Chemistry* 98, 450-461.

[93] Trifol, J., Plackett, D., Sillard, C., Szabo, P., Bras, J., Daugaard, A.E., 2016. Hybrid poly (lactic acid)/nanocellulose/nanoclay composites with synergistically enhanced barrier properties and improved thermomechanical resistance. *Polymer International* 65, 988-995.

[94] Herrera, M. A., Mathew, A. P., Oksman, K., Barrier and mechanical properties of plasticized and cross-linked nanocellulose coatings for paper packaging applications. *Cellulose* 24, 3969-3980.

[95] Chi, K., Catchmark, J. M., 2018. Improved eco-friendly barrier materials based on crystalline nanocellulose/chitosan/carboxymethyl cellulose polyelectrolyte complexes. *Food Hydrocolloids* 80, 195-205.

[96] Liu, X., Pang, J., Zhang, X., Wu, Y., Sun, R., 2013. Regenerated cellulose film with enhanced tensile strength prepared with ionic liquid 1-ethyl-3-methylimidazolium acetate (EMIMAc). *Cellulose* 20, 1391-1399.

[97] Goffin, A. L., Raquez, J. M., Duquesne, E., Siqueira, G., Habibi, Y., Dufresne, A., Dubois, P., 2011. From interfacial ring-opening polymerization to melt processing of cellulose nanowhisker-filled polylactide-based nanocomposite. *Biomacromolecules* 12, 2456-2465.

[98] Reddy, J. P., Rhim, J. W., 2014. Characterization of bionanocomposite films prepared with agar and paper-mulberry pulp nanocellulose. *Carbohydrate Polymers* 110, 480-488.

[99] Oun, A. A., Rhim, J. W., 2015. Effect of post-treatments and concentration of cotton linter cellulose nanocrystals on the properties of agar-based nanocomposite films. *Carbohydrate Polymers* 134, 20-29.

In: Cellulose to Nanocelluloses
Editor: Tri-Dung Ngo
ISBN: 978-1-53615-057-5
© 2019 Nova Science Publishers, Inc.

Chapter 7

INTERFACES IN AROMATIC POLYMER – NANOCELLULOSE COMPOSITES

Douglas M. Fox[1,*], Katrina Wakeman[1], J. I. Cruz[1], Tess Ravick[1], Whirang Cho[1], Christie Sayes[2], Jeremiah Woodcock[3] and Jeffrey W. Gilman[3]

[1]Department of Chemistry, American University, Washington, DC, US
[2]Department of Environmental Science,
Baylor University, Waco, TX, US
[3]Materials Science and Engineering Division,
Materials Measurement Laboratory,
National Institute of Standards and Technology, Gaithersburg, MD, US

ABSTRACT

Formation of well dispersed cellulose – polymer composites is a challenge. This is magnified when preparing composites using nanoscale materials, where the bulk properties are dominated by interfacial properties. In this chapter, synthetic and processing techniques for improving interfacial properties between cellulose nanocrystals and

* Corresponding Author Email: dfox@american.edu.

aromatic polymers are reviewed. A relatively new technique of exchanging sodium ions with bulk, organic cations is shown as a promising approach to improving dispersion while maintaining the potential for cellulose percolation network formation. The technique was shown to be useful in both sulfuric acid hydrolyzed cellulose nanocrystals and maleic acid hydrolyzed cellulose nanofibrils. Measurement techniques for examining the interface and associated properties dominated by the interface are discussed.

Keywords: polystyrene, bisphenol A based epoxy, glass fiber reinforced composite, cellulose nanomaterial, ion exchanged

INTRODUCTION

Most functional materials found in both nature and society are composite materials. Composites are comprised of two or more distinct materials that are combined to form a new material with improved properties. In most composites, the new material not only has combined properties from the individual components, but also has new properties not exhibited by any of the components. Composites are functional because they can be engineered to exhibit the best properties from each of the components. Further, the desirable properties of the major component (the matrix) is significantly enhanced with only a small addition of the minor component (the reinforcing filler/reinforecement). The greater the interaction between the filler and the matrix, the greater the property enhancement [1-3], although usually the ultimate properties are a compromise. In a nanocomposite, where the length scale of the filler is less than 100 nm in at least one dimension, many systems exhibit the greatest property enhancement with the smallest loading of filler.

There has been growing interest in cellulose nanomaterials (CNM) as functional materials and fillers since the turn of the century [4-6]. Cellulose is the most abundant biopolymer and is the major component in cell walls of all plants. It offers many desirable properties as a reinforcing filler in other polymers. It is readily extracted from abundant, renewable sources, constitutes 40% to 50% by mass of dry wood, nearly 90% by mass of cotton, and is even found in the exoskeleton of the Tunicate marine organism. Its solvent resistance, low abrasion, and low shrinkage make it a durable material. It is lightweight, especially compared to other fibrous fillers such as carbon, glass or steel. Therefore, its high stiffness and strength offers the opportunity to reduce weight without sacrificing performance. Cellulose can be processed into many different aspect ratios from several hundred for cellulose microfibers (CF), cellulose nanofibers (CNF) or tunicate nanocrystals (t-CNC) to 20 or less for sulfated cellulose nanocrystals (s-CNC). Depending on the morphology and dispersion of the material, a variety of optical properties, such as birefringence and transparency can be achieved. It also exhibits low gas permeability, making it a good barrier material. Pure cellulose is generally a better filler than natural fibers that contain cellulose, because the lignin and hemicellulose in natural fibers have lower thermal stability, greater variability, lower aspect ratios, poor UV stability, and lower crystallinity.

There are a number of challenges to successfully using cellulose as filler in polymeric materials [6, 7]. These challenges arise from the strong hydrogen bonding and hydrophilic nature of cellulose, and makes it incompatible with most commodity polymers, which are generally hydrophobic. Indeed, most successfully prepared cellulose – polymer nanocomposites are processed in water using hydrophilic polymers, such as polyvinyl alcohol or polyethylene oxide. Once dried, placed in a nonaqueous solvent, or mixed with a polymer melt, cellulose aggregates, and this minimizes any property enhancements. Furthermore, cellulose, especially in the form of acid hydrolyzed nanocrystals, has a lower thermal stability than most polymers and readily absorbs water. Both of these properties leads to a deterioration of interfacial adhesion and a loss in desirable composite properties, such as mechanical strength, barrier properties, and durability.

To overcome these challenges, much research has been devoted to exploring the cellulose – cellulose and cellulose – polymer interfaces [6, 8-10]. Increasing cellulose – polymer interfacial adhesion and reducing cellulose – cellulose adhesion often reduces cellulose aggregation and improves dispersion throughout the polymer. Well dispersed nanocellulose – polymer composites are often transparent. In addition, the improved stress transfer between polymer and filler greatly improves mechanical and viscoelestic properties. There are a great number of ways that the interface can be improved, including surface modification and the use of surfactants or reactive additives. There are several reviews on these methods for preparing nanocellulose – polymer composites [4, 6, 11, 12], some of which are discussed later in this chapter.

There are a variety of ways to probe the interfacial adhesion between cellulose and polymer. Increased adhesion leads to better stress transfer, culminating in improvements in mechanical and viscoelastic properties. When there is good adhesion between a filler and a polymer, stress applied to the polymer is transferred to the filler, which supports some of the load. Fibrous and other rod-like fillers, including carbon, glass, sisal, and cellulose, generally are stiffer, stronger, and more brittle than the polymer used for the matrix. As a result, good adhesion of these fillers with the polymer lead to composites that have higher modulus, higher strength, but reduced toughness compared to the pure polymer [13]. In some systems, especially in polymers that are brittle, formation of a percolation network can improve toughness. Some bulk properties are directly correlated with filler – polymer adhesion, so interface quality can be probed by measurement of these bulk properties. These include measurements in tensile, flexural, and toughness properties. Increases in tensile strength are the most often cited property. Computational analyses have shown that nanocomposite glass transition temperature (T_g) is elevated by increasing interfacial adhesion energy between the filler and matrix and maximizing the interphase volume by minimizing the average particle size [14]. Agglomeration of cellulose often leads to smaller average interparticle distances, which would reduce any increases in glass transition temperatures associated with strong particle – polymer adhesion. Tang and Weder used the percolation model of stress transfer to relate tensile storage moduli to volume fraction of cellulose participating in the stress transfer [15]. Other methods rely on visualizing the interface when stress is applied. For instance, atomic force microscopy (AFM) has been used to measure the pullout force of fibers, and in tapping mode it has been used to measure the deformation force and interphase depth moving across a polymer – filler –

polymer interface [16, 17]. There are several methods for tracking the interfacial shear stress in single fiber composites. Pull-out, microbond, fragmentation, and indentation tests can be used to monitor the interfacial shear strength [18]. Of these, the single fiber fragmentation test is generally considered the most reliable and is typically monitored using polarized microscopy. The use of mechanophores have been used recently to track stress transfer and debonding in single glass-fiber epoxy composites [19]. Finally, specifically for cellulose – polymer composites, Rusli and Eichhorn reported a method to track mechanical deformations of cellulose wihin a composite using Raman spectroscopy [20]. A theoretical model was used to provide a quantitative measure of the quality of interfacial adhesion between cellulose and polymer.

In this chapter, we briefly review cellulose nanocrystal modifications used to improve interfacial adhesion in aromatic polymer composites. We also report results using a CNM modification proceess developed in our laboratory to improve dispersion in polymer composites [7].

METHODS[1]

Materials

Methyl(triphenyl)phosphonium bromide ($MePh_3PBr$) was obtained from Alfa Aesar; sodium hydroxide (NaOH, 98.5% by mass) was obtained from Acros Organics; and hydrochloric acid (HCl, ACS Plus grade) was obtained from Fisher Scientific. Diglycidyl ether of bisphenol A epoxy resin (DGEBA) and poly(propylether)diamine (JA-d230, 230 g/mol) were obtained from Sigma Aldrich. Polystyrene (PS, 210,000 g/mol, melt flow

[1] The policy of NIST is to use metric units of measurement in all its publications, and to provide statements of uncertainty for all original measurements. In this document however, data from organizations outside NIST are shown, which may include measurements in non-metric units or measurements without uncertainty statements. The identification of any commercial product or trade name does not imply endorsement or recommendation by NIST. Opinions, interpretations, conclusions, and recommendations are those of the authors and are not necessarily endorsed by NIST.

index 7 g/10min at 200°C) was obtained from Scientific Polymer. All polymers were dried for 1 hr at 90°C in a convection oven prior to use. De-ionized water (>18.2 MΩ) was collected from an ELGA LabWater PURELAB flex 2-stage water purification system. Cellulose nanocrystals from tree pulp (NaCNC) in solution form and as a freeze-dried product were obtained from the University of Maine. These were prepared using sulfuric acid, neutralized to the sodium form, and contained 0.95% mass fraction sulfur on a dry basis. The nanocrystals had an average diameter of 6 nm and average length of 130 nm [21]. The received NaCNC solutions were diluted from 6.2 wt% mass fraction to 2.1% mass fraction using de-ionized water and the freeze-dried powder, which was dried from a solution containing a 9% mass fraction t-butanol, was used as received. Cellulose nanofibrils from tree pulp (NaCNF) were provided by the Forest Products Laboratory using maleic acid hydrolysis. Dowex 50W-X2, 50 – 100 mesh, H^+ form cation exchange resin was obtained from Sigma Aldrich. The resin was washed sequentially in 1 M HCl, deionized water, 1 M NaOH, and deionized water to remove metal impurities and convert the resin to Na^+ form.

Ion Exchange of Cellulose Nanomaterials

CNM were modified using an ion exchange process. H^+ or Na^+ form cation exchange resin was loaded into a glass chromatography column. The column was connected to a peristaltic pump and charged at a rate of 140 mL/h with a 0.1 M $MePh_3PBr$ solution until the eluent contained no more H^+ or Na^+. It was then rinsed with deionized water until the eluent contained no more surfactant molecules. NaCNC was diluted with deionized water to 4 mass-% and fed through the column at the same rate. NaCNF was fed as 1.1 mass-% solutions at the same rate. The level of exchange in the sodium cellulose was monitored in 25 mL aliquots. Portions of the collected samples were flash frozen in liquid nitrogen and freeze-dried for 2 – 3 days using a Labconco FreeZone lyophilizer. The cellulose content was determined gravimetrically from the freeze dried samples. The H^+ content was measured using a pH electrode connected to a VersaStar multimeter. The Na and P content of the solutions were determined using an Agilent 5100 inductively coupled plasma – optical emission spectrometer (ICP-OES) with Vertical Dual View configuration. All cation concentrations were measured with an uncertainty of $\sigma = \pm 0.3$ mM. A Malvern Instruments Zetasizer Nano ZS running on Zetasizer Software 7.10 was used to characterize the average particle size and surface charge of the CNM. A Surface Measurement System iGC-SEA inverse gas chromatography – surface energy analyzer was used to determine the surface area of the nanocellulose. The BET surface energy was determined using a linear regression of n-octane surface coverage between 0.05 and 1.0 fractional coverage.

Composite Processing

PS composites were prepared by extrusion in a DSM Xplore 15 mL melt mixer. Epoxy composites were prepared using a high speed planetary mixer (THINKY ARM-310 or Flack-Tek DAC 150 SP). The epoxy mixtures were poured into silicone molds, subjected to room temperature cure for at least 24 h followed by a 2 h cure at 80°C. Composites were imaged using a Zeiss ID03 inverted microscope with a CCD camera, a JEOL JSM-IT100 InTouchScope scanning electron micro-scope (SEM), and a NIST built time-correlated single-photon counting (TCSPC) instrument utilizing a femtosecond Ti-sapphire laser, as previously described [19].

Results and Discussion

Cellulose Characterization

The nanocellulose used in this study was hydrolyzed with acids that undergo secondary reaction with the anions of the acids. This produces cellulose with a charged surface. The nanofibers (CNF) in this study were formed from a carboxylic acid, producing long fibrils with carboxylic acid groups along the cellulose surface. The nanocrystals (CNC) were formed using sulfuric acid, producing short rigid rod-like crystals with sulfate half ester groups along the surface. The as received celluloses had Na^+ as the counterions. We exchanged the Na^+ in both materials to produce celluloses with $MePh_3P^+$ counterions to improve dispersion in polymer matrixes. This ion exchange process alters the surface properties of the celluloses, but does not alter the morphology. The morphologies of the nanocellulose were confirmed using SEM, as shown in Figure 1. As observed in the images, there are no significant differences in morphology after the ion exchange process.

The number of charged groups on the source CNM was determined using potentiometric titration and the counterions were determined using ICP-OES. Sodium cations, phosphonium cations, and total number of exchangeable sulfate half esters were measured using wavelengths at 589.592 nm (axial), 214.914 nm (radial), and 181.972 nm (radial), respectively. Thermal stabilities were determined by TGA under nitrogen. The apparent average size and surface charge of the CNM in aqueous solution was determined from dynamic light scattering (DLS) and zeta potential (ZP) measurements. The BET surface area was approximated using n-octane as the adsorbing gas in an iGC. These surface properties of CNM source materials and modifications are listed in Table 1.

The exchange of Na^+ is typically not complete. The more hydrophobic cations have lower affinity for the resin, generating a gradient along the column sooner than most inorganic cations. For most applications, 100% exchange is not necessary, and the high exchange ratios generated here are sufficient for improving the interface with nonaqueous solvents and polymers.

For CNC, substitution of Na^+ with organic cations generally leads to higher thermal stabilities, with the most hydrophobic cations producing CNC with the highest thermal stabilities. For CNF, substitution leads to a 2-stage decomposition, where the first step is lower than that of NaCNF, and the second step is higher.

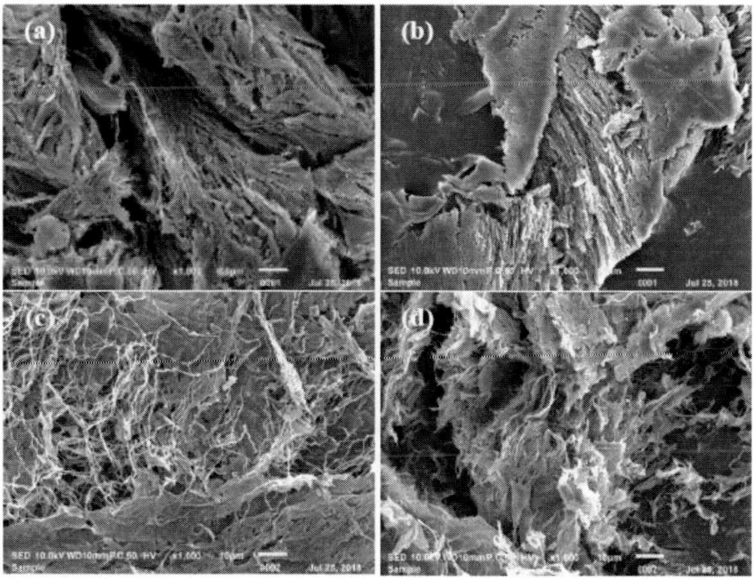

Figure 1. SEM images at 1000x magnification of (a) NaCNC, (b) $MePh_3PCNC$, (c) NaCNF, and (d) $MePh_3PCNF$.

Table 1. Characterization of Cellulose Nanomaterials

	$[Na^+]$, mM	$[M^+]$, mM	$[X^-]$, mM	$T_{0.05}$, ° C	T_{peak}, °C	σ_{avg}, nm	ZP, mV	BET, m^2/g
NaCNC	14.5	---	14.5	294	317	124	-47.4	5
$MePh_3PCNC$	1.4	6.3	7.6	291	337	124	-41.7	4
NaCNF				273	358	616	-43.9	14
$MePh_3PCNF$	0.7	6.4		233	367	417	-42.4	17

CNM are not spherical particles, so DLS provides an average apparent size of CNM materials, providing a qualitative assessment of relative sizes and aggregation in solution. The sizes of the materials are consistent with TEM, atomic force microscopy (AFM), and DLS measurements from previous studies [21-24]. The CNC exchange of Na^+ with $MePh_3P^+$ does not disrupt the colloidal stability or lead to aggregation. $MePh_3P^+$ does screen the surface charges more than Na^+, which is likely due to the large aromatic regions of the cation. The NaCNF solutions were a bit more viscous than when they were originally received, while the exchanged $MePh_3PCNF$ had the same visual appearance as when they were first prepared. This suggests that NaCNF aggregates somewhat during long term storage while the exchanged materials maintain better colloidal stability. The BET surface area is consistent with previous studies [25, 26]. Fall et al. found that the BET surface area of freeze dried CNF increases as the concentration of solution decreases [26]. This may explain the differences between CNC surface areas (which were freeze dried from solutions that were 4% cellulose by mass) and CNF surface areas (which were freeze dried from solutions that were 1% cellulose by mass). The exchange of Na^+ with $MePh_3P^+$ leads to slightly higher surface areas. The cation can act as a surfactant in water, reducing the hydrogen bonding ability of the cellulose, the surface tension of the water, and the average crystal size of the water when frozen. This may reduce the CNM aggregation during the freezing process, resulting in more void space between the cellulose chains and a larger measureable surface area.

Polystyrene Composites

The intermolecular forces between cellulose, particularly NaCNC, and polystyrene are quite different than one expects. The surface energy of cellulose is dominated by dispersion forces, especially when interacting with small molecules. These forces are from the pyranose ring structure, and are aliphatic in nature. In addition, there are strong hydrogen bonding forces due to the three hydroxyls per glucose molecule and electrostatic interactions from the sulfate half esters. Polystyrene, on the other hand, is dominated by nonpolar, aromatic interactions. The aromatic structure does allow for some hydrogen bond acceptor and cation – π characteristics, though they are rather weak. As a result, unmodified cellulose aggregates extensively when blended with polystyrene. In addition, the lower thermal stability of the ionic forms of cellulose lead to degradation of the cellulose during melt extrusion processing.

The interfacial adhesion and dispersion of cellulose has been improved by either the addition of a surfactant or chemical grafting onto the cellulose surface. Rojas et al. added sorbitan monosterate to cellulose to reduce cellulose agglomeration and improve the electrospinning processing of PS composites beading [27]. However, high levels of surfactant were needed, typically masses equal to that of the cellulose, and beading was still evident in most samples. Lin and Dufresne used PEO to improve cellulose dispersion, examining both physisorbed and chemically grafted PEG chains to the cellulose [28]. The covalent attachment of PEG did increase the thermal stability and hydrophobicity of the cellulose. Extruding the grafted cellulose with polystyrene produced composites with greater light transmittance, storage modulus, and water vapor barrier properties. However, the light transmittance improvement was minimal, the aggregation was still visually evident without magnification, the cellulose darkened in color, and there was no significant difference in glass transition. So, presumably, the composites still did not exhibit strong interfacial adhesion. Yin et al. modified carboxylated cellulose nanocrystals using a partial ion exchange procedure, replacing most of the Na^+ counterions with benzyltrimethylammonium, stearyltrimethylammonium, or diallyldimethyl-ammonium ions [29]. They prepared PS composites by solvent casting in THF. Thermal stability, glass transition temperature, Young's modulus, breaking strength, and elongation at break all improved with the modifications. The improvements indicate less agglomeration, better dispersion, and stronger interfacial adhesion.

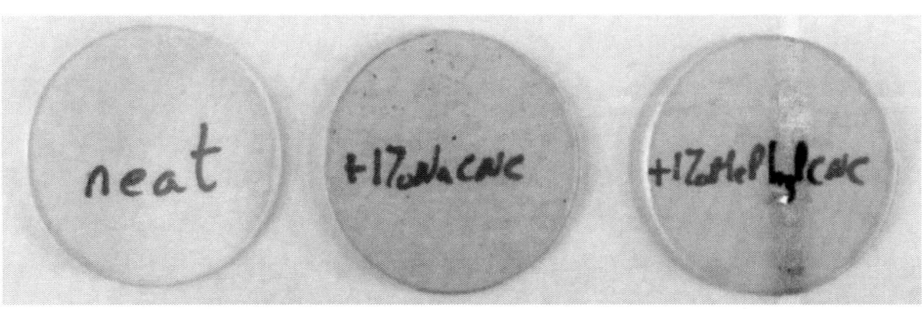

Figure 2. Extruded and pressed PS (left), 1% CNC-PS composites NaCNC (middle), and 1%$MePh_3PCNC$ (right).

Our process uses an ion exchange column and can control both the exchange efficiency and eliminate the need to isolate the modified crystals.

We found that ion exchanging Na^+ with the $MePh_3P^+$ ion increased the thermal stability and compatibility of the cellulose, which produced excellent melt blended cellulose PS nanocomposites. The effect of the extrusion and compression molding processes on the composites is shown in Figure 2. The 1%$MePh_3$PCNC sample (right) has no visible aggregates and shows no color or evidence of degredation as compared to the 1%NaCNC sample (center).

Epoxy Composites

Most common epoxy composites in use are two-component, consisting of an oligomeric-epoxide molecule (epoxy resin) and a curing agent (or hardener). The monomers can have a wide range of characteristics, from aliphatic to aromatic to hydrophilic. One of the most prevelant epoxy systems is the diglycidyl ether of bisphenol A (DGEBA) epoxy resin which typically uses an amine curing agent. Many of these are room temperature curing agents which have the polyethylene oxide structure, rendering them somewhat hydrophilic. This feature has generated interest in blending cellulose nanomaterials with epoxy composites.

Most studies to date have used unmodified NaCNC with different processing techniques employed to improve dispersion. The most common method to incorporate cellulose nanocrystals into epoxy is adding them to a more hydrophilic diamine curing agent first, with the aid of a solvent to form a well dispersed network in an emulsion, or a gel [15, 30]. The epoxy resin is then added and the solvent is removed under vacuum. The cured epoxy composite often exhibits good dispersion of the cellulose with increased stiffness. The glass transition of the composites typically decreased with NaCNC content, possibly due to incomplete cure, effects of agglomeration and weak interfacial adhesion with the epoxy. Eichhorn et al. used Raman spectroscopy to show that higher aspect ratio crystals have better adhesion and better stress transfer than shorter crystals [20]. Youngblood et al. demonstrated that adhesion could be improved, especially at lower loading levels, by choosing the appropriate curing agent [30]. The stronger adhesion appears to be related to the presence of unreacted primary amines or secondary amines [15, 30]. One of the challenges this approach has is that there is often water or solvent present, limiting cellulose – polymer interfacial adhesion and reducing the property enhancement benefits.

To improve on the adhesion between the cellulose and epoxy, surfactants [31] and cellulose surface modifications [32-35] have been used. The use of triblock poly(ethylene oxide) – poly(propylene oxide) – poly(ethylene oxide) nonionic surfactants reduced aggregation and moisture absorption of the NaCNC, but did not significantly improve adhesion with the epoxy matrix. As a result, only tensile modulus increased, with glass transition temperatures, tensile strength, and tensile strain all decreasing. Esterification using anhydrides has been successfully employed to reduce aggregation, increase matrix adhesion, and enhance mechanical properties [32-34]. Some of the modifications even increased toughness while improving strength, which is a remarkable achievement. However, there are indications that residual solvent and alkyl chain mobility diminishes the reinforcing effects of the cellulose and the added processing required to modify the crystals reduces the cost effectiveness of using cellulose reinforcements.

Ion exchanging CNMs greatly improves the processability of the composites and maintains the ability to form percolated interdigitated cellulose networks. Ion exchanged-CNMs can be successfully blended with one of the epoxy components using a high speed mixer, without the use of solvent or additives. As shown in Figure 3, using NaCNCs causes aggregation in the composites and the NaCNC degrades when heated at 105°C for 24 h. However, the CNCs modified with $MePh_3P^+$ are well dispersed and do not darken or decompose when held at elevated tmperatures. As a result, the $MePh_3PCNC$ containing epoxies absorb less water, are stiffer, and have higher tensile strengths than NaCNC containing ones [7]. The driving force appears to be due to significant reductions in cellulose – cellulose interactions, rather than increased cellulose – epoxy adhesion. As shown in Figure 4, the improvements in dispersion are evident at all length scales. The average CNM particle size is 2 orders of magnitude smaller in area and CNM particles are more needle-like in structure, increasing the aspect ratio of the particles. The modification procedure can also be used with larger aspect ratio CNM, such as the CNF produced from maleic acid hydrolysis, which shows the versatility of this process.

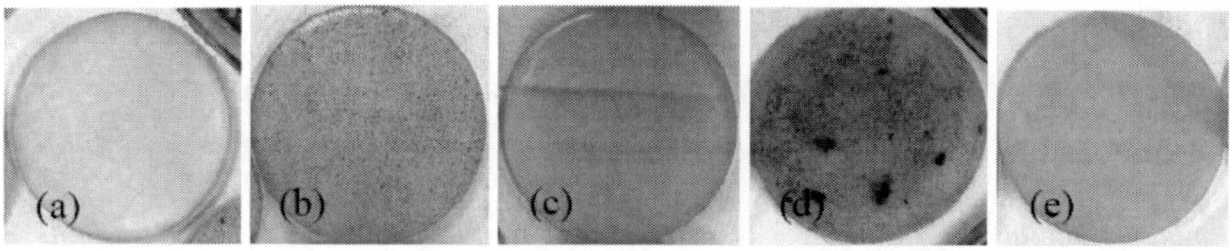

Figure 3. Epoxy composites exposed to water for 30 days and dried in a convection oven at 105°C for 24 h containing (a) no cellulose, (b) 1 mass-% NaCNC, (c) 1 mass-% $MePh_3PCNC$, (d) 5 mass-% NaCNC, and (e) 5 mass-% $MePh_3PCNC$.

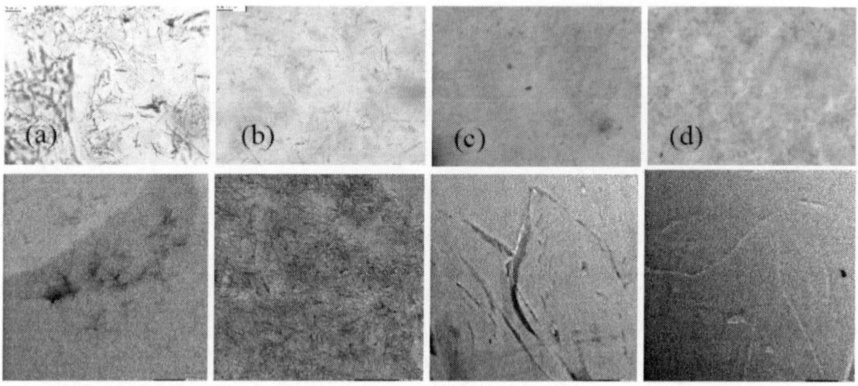

Figure 4. 10x optical microscope (top row), and transmission electron microscope images (bottom row) of epoxy composites containing (a) 5% NaCNC, (b) 5% $MePh_3PCNC$, (c) 5% NaCNF, and (d) 5% $MePh_3PCNF$.

One of the potential applications of using cellulose – epoxy nanocomposites is in lightweighting the composites. This extends to fiber reinforced plastics (FRP), especially if some of the traditional glass fiber or basalt fibers can be removed by the addition of cellulose. In an extended study, Kyriaki et al. has found that NaCNC can increase the fiber – epoxy adhesion in glass fiber filled epoxy using a sheet molding compounding (SMC) technique [36-38]. The NaCNC were introduced either by adding directly to the hardener [36] or coating the glass surface by solution immersion [37]. In both cases, modulus, strength, and ultimate strain, in both the tensile and flexural configurations, were increased by the addition of NaCNC, with the maximum benefits occurring around 1% to 1.5% mass fraction loading. The tensile and flexural properties of composites containing 35 mass-% glass fibers could be replicated by using 25 mass-% glass fiber plus 1–1.5 mass-% NaCNC, and this resulted in a 7.5% reduction in the mass of the composite [38].

CONCLUSION

The interface between cellulose nanocrystals and the polymer matrix is key to many composite properties. Improvements in adhesion lead to improved dispersion and enhanced glass transition temperatures, strength, and in some cases toughness. In this chapter we reported on the use of methyl(triphenyl)phosphonium cation exchanged cellulose nanomaterials to improve the interfacial properties between cellulose and aromatic polymers. In composites containing $MePh_3P^+$ exchanged CNM, fewer aggregates were observed, less water was absorbed, and no degradation was observed as compared to the NaCNC composite. The cellulose – cellulose interactions in the exchanged celluloses were significantly lower than in sodium celluloses without much loss in the cellulose – polymer adhesion. As a result, dispersion was favored over aggregation.

ACKNOWLEDGMENTS

This work was supported in part by NSF (Grant No. CHE 1625977) and Air Force Office of Scientific Research (Award No. F1ATA00236G002).

References

[1] Krishnamoorti, R., Vaia, R. A., (Eds.), 2001. *Polymer Nanocomposites: Synthesis, Characterization, and Modeling.*

[2] Mai, Y.-W., Yu, Z.-Z., (Eds.), 2006. *Polymer Nanocomposites*, CRC Press, Boca Raton.

[3] Young, R. J., Lovell, P. A., 2011. *Introduction to Polymers*, 3rd ed., CRC Press, Boca Raton.

[4] Hubbe, M. A., Rojas, O. J., Lucia, L. A., Sain, M., 2008.) *Cellulosic Nanocomposites: A Review, BioResources* 3, 929-980.

[5] Dufresne, A., 2013. Nanocellulose: a new ageless bionanomaterial, *Materials Today* 16, 220-227.

[6] Mariano, M., El Kissi, N., Dufresne, A., 2014. Cellulose nanocrystals and related nanocomposites: Review of some properties and challenges, *Journal of Polymer Science Part B: Polymer Physics* 52, 791-806.

[7] Fox, D. M., Rodriguez, R. S., Devilbiss, M. N., Woodcock, J., Davis, C. S., Sinko, R., Keten, S., Gilman, J. W., 2016. Simultaneously Tailoring Surface Energies and Thermal Stabilities of Cellulose Nanocrystals Using Ion Exchange: Effects on Polymer Composite Properties for Transportation, Infrastructure, and Renewable Energy Applications, *ACS Applied Materials & Interfaces* 8, 27270-27281.

[8] Renneckar, S., Zink-Sharp, A., Esker, A. R., Johnson, R. K., Glasser, W. G., 2006. Novel methods for interfacial modification of cellulose-reinforced composites, In *Cellulose Nanocomposites: Processing, Characterization, and Properties* (Oksman, K., and Sain, M., Eds.), pp 78-96.

[9] Cherian, B. M., Leao, A. L., de Souza, S. F., Thomas, S., Pothan, L. A., Kottaisamy, M., 2011. Cellulose Nanocomposites for High-Performance Applications, In *Cellulose Fibers: Bio- and Nano-Polymer Composites: Green Chemistry and Technology* (Kalia, S., Kaith, B. S., and Kaur, I., Eds.), pp 539-587, Springer Berlin Heidelberg, Berlin, Heidelberg.

[10] Miao, C. W., Hamad, W. Y., 2013. Cellulose reinforced polymer composites and nanocomposites: a critical review, *Cellulose* 20, 2221-2262.

[11] Ramires, E. C., Dufresne, A., 2011. A review of cellulose nanocrystals and nanocomposites, *Tappi Journal* 10, 9-16.

[12] Aitomaki, Y., Oksman, K., 2014. Reinforcing Efficiency of Nanocelluloses in Polymer Nanocomposites, In *Handbook of Green Materials, Vol 2: Bionanocomposites: Processing, Characterization and Properties* (Oksman, K., Mathew, A. P., Bismarck, A., Rojas, O., Sain, M., and Qvintus, P., Eds.), pp 131-145.

[13] Oksman, K., Clemons, C., 1998. Mechanical properties and morphology of impact modified polypropylene - Wood flour composites, *J. Appl. Polym. Sci.* 67, 1503-1513.

[14] Sinko, R., Qin, X., Keten, S., 2015. Interfacial mechanics of cellulose nanocrystals, *Mrs Bulletin* 40, 340-348.

[15] Tang, L. M., Weder, C., 2010. Cellulose Whisker/Epoxy Resin Nanocomposites, *ACS Applied Materials & Interfaces* 2, 1073-1080.

[16] Li, Y., Huang, Y., Krentz, T., Natarajan, B., Neely, T., Schadler, L. S., 2017. Polymer Nanocomposite Interfaces: The Hidden Lever for Optimizing Performance in Spherical Nanofilled Polymers, In *Interface/Interphase in Polymer Nanocomposites* (Netravali, A. N., and Mittal, K. L., Eds.), pp 3-69.

[17] Liu, L., 2017. Investigation of Interfacial Interactions between Nanofillers and Polymer Matrices Using a Variety of Techniques, In *Interface/Interphase in Polymer Nanocomposites* (Netravali, A. N., and Mittal, K. L., Eds.), pp 255-282.

[18] Moon, C. K., Holmes, G. A., McDononough. W. G., 2007. The effect of inter-fiber distance on the interfacial properties in E-glass fiber/epoxy model composites, *J. Appl. Polym. Sci.* 105, 3483-3491.

[19] Woodcock, J. W., Beams, R., Davis, C. S., Chen, N., Stranick, S. J., Shah, D. U., Vollrath, F., Gilman, J. W., 2017. Observation of Interfacial Damage in a Silk-Epoxy Composite, Using a Simple Mechanoresponsive Fluorescent Probe, *Advanced Materials Interfaces 4*, 1601018.

[20] Rusli, R., Shanmuganathan, K., Rowan, S. J., Weder, C., Eichhorn, S. J., 2011. Stress Transfer in Cellulose Nanowhisker Composites—Influence of Whisker Aspect Ratio and Surface Charge, *Biomacromolecules* 12, 1363-1369.

[21] Sacui, I. A., Nieuwendaal, R. C., Burnett, D. J., Stranick, S. J., Jorfi, M., Weder, C., Foster, E. J., Olsson, R. T., Gilman, J. W., 2014. Comparison of the Properties of Cellulose Nanocrystals and Cellulose Nanofibrils Isolated from Bacteria, Tunicate, and Wood Processed Using Acid, Enzymatic, Mechanical, and Oxidative Methods, *ACS Applied Materials & Interfaces* 6, 6127-6138.

[22] Beck, S., Bouchard, J., Berry, R., 2012 Dispersibility in Water of Dried Nanocrystalline Cellulose, *Biomacromolecules* 13, 1486-1494.

[23] Peng, Y., Gardner, D. J., Han, Y., 2012. Drying cellulose nanofibrils: in search of a suitable method, *Cellulose* 19, 91-102.

[24] Chang, H., Luo, J., Bakhtiary Davijani, A. A., Chien, A.-T., Wang, P.-H., Liu, H. C., Kumar, S., 2016. Individually Dispersed Wood-Based Cellulose Nanocrystals, *ACS Applied Materials & Interfaces* 8, 5768-5771.

[25] Khoshkava, V., Kamal, M. R., 2014. Effect of drying conditions on cellulose nanocrystal (CNC) agglomerate porosity and dispersibility in polymer nanocomposites, *Powder Technology* 261, 288-298.

[26] Munier, P., Gordeyeva, K., Bergstrom, L., Fall, A. B., 2016. Directional Freezing of Nanocellulose Dispersions Aligns the Rod-Like Particles and Produces Low-Density and Robust Particle Networks, *Biomacromolecules* 17, 1875-1881.

[27] Rojas, O., Montero, G., Habibi, Y., 2009. *Electrospun Nanocomposites From Polystyrene Loaded With Cellulose Nanowhiskers*, Vol. 113.

[28] Lin, N., Dufresne, A., 2013. Physical and/or Chemical Compatibilization of Extruded Cellulose Nanocrystal Reinforced Polystyrene Nanocomposites, *Macromolecules* 46, 5570-5583.

[29] Yin, Y., Hong, Z., Tian, X., Zhu, Q., Jiang, X., Wang, H., Gao, W., 2017. Cellulose nanocrystals modified with quaternary ammonium salts and its reinforcement of polystyrene, *Polymer Bulletin*.

[30] Peng, S. X., Moon, R. J., Youngblood, J. P., 2014. Design and characterization of cellulose nanocrystal-enhanced epoxy hardeners, *Green Materials* 2, 193-205.

[31] Emami, Z., Meng, Q., Pircheraghi, G., Manas-Zloczower, I., 2015. Use of surfactants in cellulose nanowhisker/epoxy nanocomposites: effect on filler dispersion and system properties, *Cellulose* 22, 3161-3176.

[32] Wang, W.-J., Wang, W.-W., Shao, Z.-Q., 2014. Surface modification of cellulose nanowhiskers for application in thermosetting epoxy polymers, *Cellulose* 21, 2529-2538.

[33] Abraham, E., Kam, D., Nevo, Y., Slattegard, R., Rivkin, A., Lapidot, S., Shoseyov, O., 2016. Highly Modified Cellulose Nanocrystals and Formation of Epoxy-Nanocrystalline Cellulose (CNC) Nanocomposites, *ACS Applied Materials & Interfaces* 8, 28086-28095.

[34] Peng, S. X., Shrestha, S., Yoo, Y., Youngblood, J. P., 2017. Enhanced dispersion and properties of a two-component epoxy nanocomposite using surface modified cellulose nanocrystals, *Polymer* 112, 359-368.

[35] Yue, L., Maiorana, A., Khelifa, F., Patel, A., Raquez, J.-M., Bonnaud, L., Gross, R., Dubois, P., Manas-Zloczower, I., 2018. Surface-modified cellulose nanocrystals for biobased epoxy nanocomposites, *Polymer* 134, 155-162.

[36] Asadi, A., Miller, M., Sultana, S., Moon, R. J., Kalaitzidou, K., 2016. Introducing cellulose nanocrystals in sheet molding compounds (SMC), *Composites Part A: Applied Science and Manufacturing* 88, 206-215.

[37] Asadi, A., Miller, M., Moon, R. J., Kalaitzidou, K., 2016. Improving the interfacial and mechanical properties of short glass fiber/epoxy composites by coating the glass fibers with cellulose nanocrystals, *Express Polymer Letters* 10, 587-597.

[38] Asadi, A., Miller, M., Singh, A. V., Moon, R. J., Kalaitzidou, K., 2017. Lightweight sheet molding compound (SMC) composites containing cellulose nanocrystals, *Composite Structures* 160, 211-219.

In: Cellulose to Nanocelluloses
Editor: Tri-Dung Ngo
ISBN: 978-1-53615-057-5
© 2019 Nova Science Publishers, Inc.

Chapter 8

BARRIER PROPERTIES OF NANOCELLULOSE BASED FILMS

Rehan Ali Pradhan, Muhammad Zubair, Muhammad Safder, Muhammad Arshad and Aman Ullah*

Department of Agriclutural, Food and Nutritional Science, University of Alberta, Edmonton, Alberta

ABSTRACT

Advancements in technology have focused on food packaging as a result of environmental problems, consumer health and economic restrictions. Nanoellulose has unique properties due to its easy availalibliy, renewability and biodegradability. These promising characteristics make it a suitable nanofiller to develop low cost, eco-friendly and lightweight films or coatings for food packaging applications. A synoptic view of cellulose nanocrystal (CNC) and cellulose nanofibre (CNF) based sustainable and green films or coatings is given with emphasis on barrier properties. This chapter provides a comprehensive overview of the current advances in mositure and oxygen barrier properites of nanocellulose based coatings and

* Corresponding Author Email: amanullah@ualberta.ca.

films. The chapter also covers the brief description of molecule transport mechanism through films or coatings and factors affecting the barrier properties. Lastly, we discuss our viewpoints on future research of CNC/CNF derived films and coatings to substitute synthetic materials with new cellulose bionanocomposites.

Keywords: nanocellulose, composites, food packaging, barrier properties

INTRODUCTION

Packaging is an effective way to maintain the quality of the products until their end use. It acts as a barrier against physical damages, biochemical and microbial spoilage in various industries includes food and beverages, cosmetics, pharmaceuticals and other consumer goods. Food packaging is one of the fastest growing industries in the world with an annual growth rate of 4.6% [1, 2]. In food industry, packaging offers barrier against water vapor, odor, grease, oxygen, microorganisms etc. [3, 4]. Nowadays, most of the food packaging materials are derived from tin, aluminum, glass and fossil fuel based synthetic polymers. These packaging materials are very strong and have excellent barrier properties [5, 6]. However, some of these materials are from non-renewable and non-degradable resources. The utilization of bio-based materials as a potential substitute of petroleum derived food packaging materials have attracted much attentions from academia and industry due to their environment friendly nature and safe disposal [6].

Cellulose nanomaterials are one of the viable choices to address issues related to petro-based polymers for food packaging application [7]. As cellulose is the naturally occurring most abundant biopolymer and is around 40% of lignocellulosic biomass [8]. Cellulose derived material is cheap, lightweight, sustainable and biodegradable [9]. But, there are various issues with cellulose based material because of their poor barrier properties, particularly for grease, oxygen and water [10]. However, cellulose nanocrystals (CNCs) and cellulose nanofibrils (CNFs) production from cellulose have unfold new horizons to exploit the cellulose-based materials for food packaging applications with improved barrier properties. Nanocellulose based bionanocomposites have large surface with a small diameter ranging between 2–50 nm [11, 12]. Hydrogen bonding nature of CNC provides a strong dense network responsible for hindering the transportation of gaseous molecules (water and oxygen) through the films or coating [13].

This chapter focuses on the oxygen and water vapor barrier properties of films or coating derived from nanocellulose. The mechanistic view of gas transportation is described briefly along with factors affecting barrier properties. A highlight on packaging applications is also provided.

Transport Mechanism of Molecules Through Nanocellulose Based Films or Coatings

The passage of gaseous molecules between the two phases separated through a polymeric coatings or films is a complex process that can be divided into three stages. The first stage involves adsorption of molecules onto the sample surface. In the second stage, diffusion of these molecules through the films or coating occurs. Lastly, desorption of diffused molecules from the sample surface takes place thus they exit the surface [5, 14].

During the adsorption of gas molecules on an amorphous polymeric coatings and films, they positioned themselves in the available free spaces due to Brownian motion of polymeric chains. On the other hand, diffusion of gaseous molecules occurs because of the free spaces between the phases. Consequently, number and size of free volume spaces determines diffusion through the polymer. The diffusion of gas is dependent on the available dynamic free volume because of segmental motion present in polymer chains. Whereas interaction between gaseous molecules and polymeric layers control gas solubility rate [1, 14-16].

The gas permeability through the nanocellulose based films is primarily depends on dissolution and diffusion rate of gas molecules in the films or coatings [17]. Hence, gas permeability in nanocellulose film can be stated as:

$$P = DS$$

where P = permeability, D = diffusion coefficient, and S = solubility coefficient.

The permeability coefficient P is defined by Henry's law of solubility to Fick's law of diffusion,

$$P = DS = ql/At\Delta P$$

Here q= amount of material pass through the film, l = thickness, A = cross sectional area, t = time, and Δp = pressure difference between the two sides of film.

The improvement in gas barrier properties of bionanocomposites materials is attributed to the addition of nanoparticles and has been described by two different phenomena. Firstly, the addition of nanoparticles decreases polymer chain mobility and flexibility. As a result, regularity of chain arrangement in polymer matrix increases and penetrant mobility decreases in nanocomposites as a results change in the diffusibility rate happens [18-20]. Secondly, impermeable nanoparticles in polymer formulations increase the tortuosity of the diffusive path for a penetrant molecule [21-23]. Conversely, the addition of micro-additive into the polymeric matrix is not effective to improve the barrier properties of the polymers as illustrated in Figure 1.

The improvement in gas barrier properties of nanocellulose derived bionanocomposites can be explained on the basis of nanofibrils morphology which has uniform and nano size dimensions. Thus, nanofibrils form complex dense network that enhances the tortuosity within the film. The gas permeability through the film is reduced due to this phenomenon [13]. Furthermore, nanofibrils and nanocrystals have highly crystalline structure which also contributes to act as a barrier for various gas molecules [17, 24].

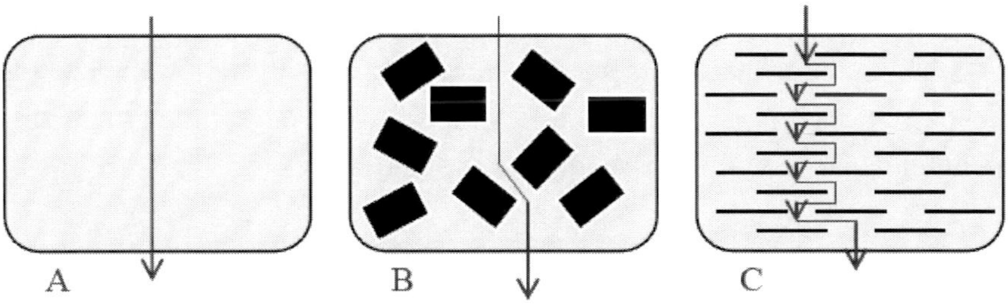

Figure 1. Gas transport via (a) pure polymer, (b) polymer with addition of micro additive, (c) increased diffusion path in the nanocellulose based films.

CNCs possesses higher crystallinity compared to CNFs, as during strong acidic hydrolysis of cellulose, crystalline parts of CNCs remains unaffected [25, 26]. Studies found that CNCs and CNFs have similar solubility, but the oxygen migrated more slowly in the CNFs film which is because of structural organization within the CNF based films responsible for low oxygen permeability. Additionally, the higher entanglements enhanced the tortuosity or the diffusion path within the film as a result oxygen transport is lowered [27]. CNCs based films have low water vapor barrier properties than CNF films. The high affinity between water and the CNCs molecules improved its water barrier properties. Due to small size and limited swelling of CNCs, it forms rigid network which has a negative effect on water diffusion. However, at high relative humidity (RH) this complex structural network can be broken down because of swelling. As a result, oxygen and water vapor barrier properties will be lowered [5, 28].

General Factors Influencing Barrier Properties of Nanocellulose Based Films or Coatings

The transfer of gaseous molecules through coatings or films mainly depends on the thermodynamic interactions between the polymer and nanoparticles. Pressure, temperature and concentration play a vital role to establish favorable enthalpy interaction between them which in turn determines the barrier properties of the bionanocomposites [29, 30]. Gas permeation rate depends on factor such as gaseous molecules diameter, polymer structural arrangement and morphology as well as film/coating thickness. Following are the factors which governs the gas permeability through the nanocellulose based bionanocomposite.

a) CNCs and CNFs geometrical shape, size and aspect ratio
b) Amount of CNCs or CNFs used in the polymer matrix
c) Homogeneity and degree of dispersion of nanocellulose within the polymer matrix
d) Orientation in polymer matrix
e) Polar character of each nanocellulose and polymer
f) Composition of the polymer
g) Bubbles or pinhole present in the films or coatings

Exfoliation of CNCs and CNFs within the polymer matrix required a compatibility between the polymer and nanoparticle surface. Also, bionanocomposites composition and permeant nature affects the gas permeability through the films or coatings [20, 31-33].

BARRIER PROPERTIES OF NANOCELLULOSE REINFORCED POLYMER-COMPOSITES

Vapor Barrier Properties

In this section, efforts have been made to discuss various achievements in vapor barrier properties of nanocellulose based bio-composites through use of different polymer matrix such as modified cellulose esters, chitosan, poly(3-hydroxybutyrate) PHB, whey protein isolate, pectin and others. Various testing methods have also been mentioned to avoid the variation while comparing between different matrix systems. As discussed in our previous sections, cellulose contains multiple hydroxyl groups which makes it hydrophilic in nature as a results moisture adsorption take place [34].

Swelling of CNF films in higher humidity also result an increase in permeation values. Water molecules penetrate the cellulose fibrils resulting in breaking of the hydrogen bonds and fragmentation of long chain held together, thus restricting their potential applications. It is widely accepted that high water repellence properties can be obtained by covering cellulose film surfaces by hydrophobic components [35]. Using the underlying principle, significant improvement on vapor barrier properties was achieved by incorporating nonpolar hydrocarbon side-chain in cellulose ester coatings. Three layered film (ester-CNF-ester) was prepared using cellulose ester with different even numbered chain lengths (C6 to C18) of fatty acid chloride and CNF [36]. Willberg et al. produced CNF by feeding suspension of pre-refined fibre through microfluidics fluidizer. Enhanced vapor barrier characteristics was achieved, and wettability of CNF film was also improved through corona treatment. ASTM-E–96 B procedure which was slightly modified was followed to determine water vapor transmission rates (WVTR) through the CNF films. The study demonstrated the decrease in moisture permeability with increase in length of non-polar carbon chain. By using C18 coating, water vapor transmission rates decreased by approximately 50% to 300 $g/m^2/day$ from 650 $g/m^2/day$.

In the study by Hu et al. rice straw was used as biomass to produce all lignocellulosic-composite films. Nanocellulose [CNC, alkaline CNF (ACNF), oxidized CNC(OCNC) and aqueous counter collision generated CNFs (ACCNF)] with four different aspect ratios and lateral dimensions were isolated. The addition of ACNF (10 wt.%) demonstrated increase in WVTR due to its hygroscopic nature. Increasing the ACNF content to 20 wt.% and 30 wt.% decreased the permeability values, suggesting the dominance of crystalline effects. ACCNF exhibited lowest moisture affinity as shown in Figure 2 [37].

Aulin et al. achieved significant improvement in moisture barrier property of nanocellulose precoated sheets by depositing alkyd resins (used as moisture sealant) [38]. Alkyd resins are polymer of esters where monomer entity is composed of fatty acids or alcohols containing two or more hydroxyl groups[39]. The acceptable WVTR upper limit at 25°C and 50% relative humidity for a material to come under the category of "high moisture barrier" is 5 g/m^2.day [38, 40]. It was observed that moisture barrier property not only depends on chain length of fatty acids but also on the degree of unsaturation due to formation of more crosslinked structure. For same oil length, linseed oil (composed of 50% linolenic acid which contains 3 unsaturated double bonds) was found to be more efficient in reducing WVTR as compared to tall oil (which contains 50% of oleic moiety having one unsaturated double bond and 35% of linoleic moiety which is composed of two unsaturated double bonds). Substrate coated with linseed oil-based alkyd resin/nanocellulose exhibited WVTR of 2.0 g/m^2.day and tall oil-based alkyd resin/nanocellulose exhibited transmission rate of 4.2 g/m^2.day. It was further reduced to 0.8 g/m^2.day for linseed oil-based alkyd resin and 3.1 g/m^2.day for tall oil-based alkyd resin, by applying 6 g/m^2 nanocellulose prior to alkyd resins coat. It exhibited the lowest WVTR for the reported values for bio-based coatings at standard conditions of 23°C temperature and relative humidity of 50%.

Rafieian et al. prepared a nanocomposite film for packaging application comprising of gluten and carboxylated CNCs [41]. An oxidation reaction was employed to convert -OH group at C6 position of cellulose into carboxyl groups. This was performed to obtain repulsion between individual molecular chains and thus avoid agglomeration of CNCs filler. This helped to achieve stable dispersion of cellulose nanocrystals in aqueous solution. For carboxylated CNCs loadings above 2.5wt.%, the water vapor permeability (WVP) decreased significantly for subsequent loadings of 5 and 7.5wt.%. This was attributed to the increased tortuous path encountered when vapor molecules diffused through the packaging film. The WVP decreased by 36% (4.96×10^{-11} mol/m.s. Pa) with 7.5 wt.% carboxylated CNCs. The increase in WVP for the films composed of 10 wt.% carboxylated CNCs was likely due to less tortuous path once the agglomeration of carboxylated CNCs takes place [41].Nanocellulose reinforced methycellulose-based packaging film was prepared by Khan et al. Significant reduction in water vapor permeability was achieved with low loadings of CNC (0-1wt.%) as shown in Figure 3. [42].

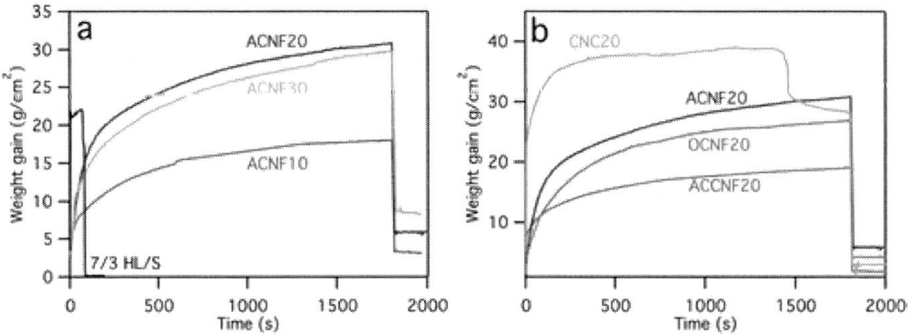

Figure 2. Water sorption of composite films reinforced with (a) 10 wt.%, 20 wt.% and 30 wt.% ACNF (b) 20 wt.% CNC, ACNF, OCNF and ACCNF. Reproduced with permission from ACS [37].

In another study, Pereira et al. used citric acid (0–30 wt.% based on hemicellulose) along with nanocrystals of cellulose (CNCs 0-8 wt.% based on hemicellulose) to prepare wheat straw derived hemicellulose based biodegradable films. E96-05 method was employed to measure barrier properties.[8] Citric acid showed both plasticizing as well as crosslinking effects. The optimized value was obtained with 5.9 wt.% CNCs and citric acid amount of 30 wt.% as a results lowest water permeability values (1.68 g mm h^{-1} kPa^{-1} m^{-2}) were obtained [43].

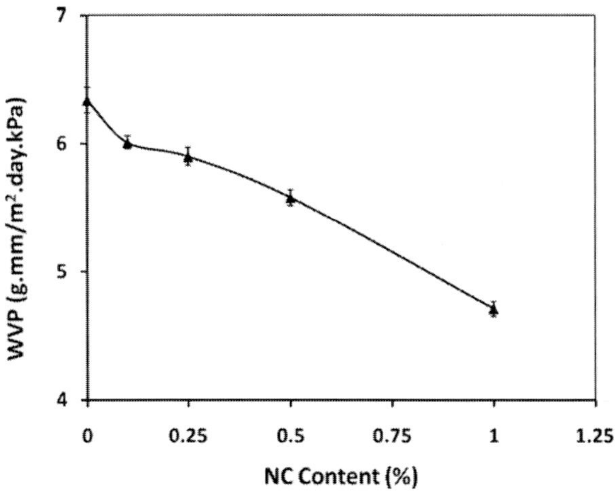

Figure 3. Water vapor permeability of methylcellulose-based biodegradable film reinforced with CNC(0-1 wt.%). Reproduced with permission from ACS [42].

Cherpinski et al. investigated the effect of electrospun PHB and PHBV [poly(3-hydroxybutyrate) and poly(3-hydroxybutyrate-co-3-hydroxyvalerate)] coatings on the barrier properties of CNF films. The permeability values were determined using the gravimetric method of ASTM 2011. Water vapor transport properties (water vapor permeance) was measured as 3.7×10^{-9} $kg/m^2s^{-1}Pa^{-1}$ for the CNF film. Significant improvement was obtained for electrospun PHB films coated on both sides with nano papers. The values obtained were in the range of $2.5–3.5 \times 10^{-11}$ $kg/m^2\ s^{-1}\ Pa^{-1}$ [44].

Chi et al. demonstrated production of ternary complex consisting of cross-linked chitosan and carboxymethyl cellulose as a continuous phase and cellulose nanocrystal employed as a reinforcement [45]. The WVTR of the film was determined gravimetrically following the ASTM E 96-95 procedure with slight modification [46]. The WVTR was normalized by considering the sample thickness. Calculations were made based on the formula given below:

WVTR ($g\ \mu m\ m^{-2}\ d^{-1}$) = (mass of water reduction×sample thickness)/(time×area) [46].

Films obtained with 10 wt.% CNCs contents exhibited water vapor transportation rate value of 7982 $g\ \mu m\ m^{-2}\ d^{-1}$. No further improvement was achieved with the increase of CNCs contents which may be due to CNCs aggregation.

Qazanfarzadeh et al. used solution casting method to prepare composite films from whey protein isolate and oat husk nanocellulose [47]. The whey protein films have moderate barrier properties as they comprise of hydrophilic amino acids. Oat husk nanocellulose was incorporated to improve water vapor barrier characteristics of these films. The water vapor permeability of whey protein-based composite films was measured using the standard method of ASTM E96-93. The permeability of control film was calculated as 5.256×10^{-8}g m/Pa h m^2, which reduced significantly by 14%, 34% and 36% by incorporating 2.5, 5 and 7.5 wt.% of oat husk nanocellulose respectively.

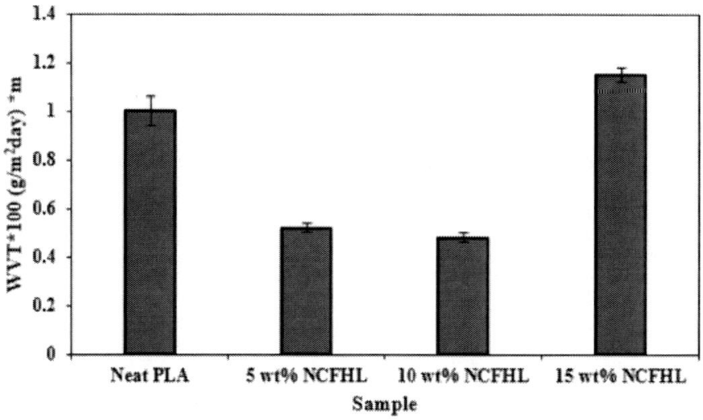

Figure 4. WVTR of PLA/NCFHL biocomposites. Reproduced with permission from ACS [48].

Polylactic acid (PLA) bio-composites were prepared by addition of nanocellulose fibrils with high lignin content (NCFHL) [48]. The water vapor transmission was reduced to half with the addition of 10 wt.% NCFHL. Further addition of nanofiller yielded poor barrier characteristics as shown in Figure 4.

In another study by Chaichi et al. the edible pectin film was incorporated with three fractions of CNCs (2%, 5% and 7% w/w). Films prepared by solution casting method were investigated for water vapor barrier properties by ASTM E96 method. Reinforcement of pristine with 5% CNCs results in the reduction of water vapor permeability (WVP) from 1.28×10^{-10} g/m s Pa to 9.3×10^{-11} g/m s Pa. Aggregation was observed above 7% concentration of CNCs. Trifol et al. used combination of nanocellulose (1 and 5 wt.%) and nanoclay (1, 3 and 5 wt. %), to investigate the barrier properties of the composites. Polyethylene terephthalate (PET), which is accepted as the industrial standard in terms of permeability values has WVTR of around 420 g μm m^{-2} day^{-1} at temperature of 23°C and 85% relative humidity, while PLA shows a WVTR of 5250 g μm m^{-2} day^{-1} at the same temperature and relative humidity of 90%. More reduction is therefore required for PLA to successfully compete with the barrier properties of PET [49]. It was found that the addition of clay with nanocellulose resulted in synergistic performance through a reduction in the water vapour transmission rate of up to 76%.

The study by Silva et al. demonstrated the effectiveness of the edible coating prepared from cellulose nanofibers, hydroxypropyl methyl cellulose, glycerol, and k-carrageenan [50]. Reduction of approximately 30% in final weight loss was shown for coated grapes and it also resulted in the improvement of permeability values by 34%. The measurement for WVTR was carried out using the method E96/E96M (ASTM). The coated grapes demonstrated significant reduction in WVP values (2.5×10^{-14}) as compared to control uncoated sample (3.8×10^{-14}). The study thus suggested an improvement in the shelf life for coated grapes by using the edible coating.

Mirmehdi et al. coated paper substrate with CNFs suspensions. The measurement for WVTR was performed using the method E96/E96M (ASTM), standard temperature of 23°C and 50% relative humidity. This suspension decreased WVTR of the paper from 28.5 g /m^2.day to 24 g /m^2. day using 11μm layer thicknessof CNFs. This showed 16% decrease when compared to control paper [51].

Nanocellulose (NC) functionalized coating obtained by layer by layer technique was examined by Herrera et al. [52]. Citric acid crosslinker (X) was incorporated to enhance the moisture performance of these coatings. The measurement for WVTR of coated and uncoated samples were performed using the modified method E96/E96M (ASTM). Coated samples showed lower WVTR than the control sample. There was reduction in water vapor permeability values by 60%.

PLA multilayer films with alternating layers of nano-fibrillated cellulose (NFC) or carboxymethyl cellulose (CMC) were prepared by layer-by-layer deposition method using polyethyleneimine (PEI) as a polyelectrolyte [55]. A 20 bilayer PEI/NFC showed reduction in water permeance values by 30%. It was reduced drastically by 50% for a 50 bilayer PEI/NFC as compared to uncoated film. The barrier performance of NFC coated PLA films were comparatively better than CMC counterpart as shown in Figure 5 [53].

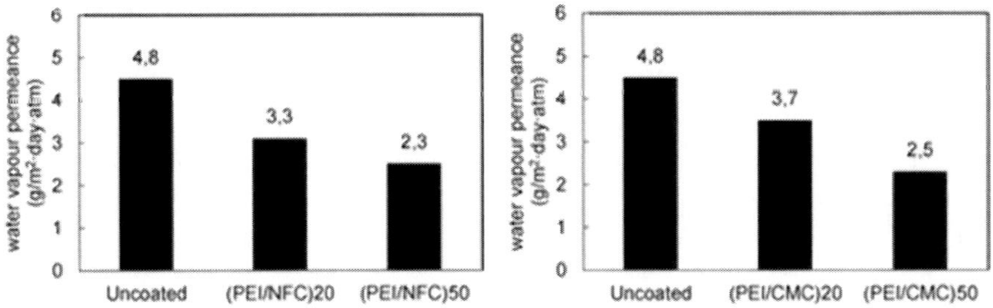

Figure 5. Water vapor permeability of PLA films coated with PEI/NFC and PEI/CMC. Reproduced with permission from ACS [53].

Oxygen Barrier Properties

Different synthetic polymers i.e., polyethylene (PE), and polypropylene (PP) have been used for packaging of food and medicines. Wide use of these synthetic polymers was due to high strength, low cost of production, chemical stability and viscoelastic properties. Most of these polymer sheets were oxygen permeable with oxygen transmission range (OTR) of 40-1000cc/m2/day [54]. Limited flow of oxygen is required to reduce rate of oxidation of stored material. Different types of halogen and metal films were synthesized to reduce the OTR 0.1-10cc/m2/day [54]. These polymers had some drawbacks associated health and environmental problems. Such as incineration of polyvinylidene chloride (PVDC) produce dust and dioxins. Metal polymers are considered non-recyclable materials [55].

As shown in Figure 6, the nanocellulose based films were produced from birch pulp and it showed excellent oxygen barrier as low as 0.12 cm^3 μm/(m^2 d kPa) at 50% relative humidity [56].

Cellulose nanofibers sheet owes some characteristic properties which make its use diverse. CNFs sheets regulate the flow of air (oxygen) that is required for packing and storing of food and medicine. CNFs is sustainable and biocompatible material without health and environmental impacts. It can be produced easily by mechanical disintegrating highly crystalline nanofibrils in cellulose bulk, the most abundant biomass. CNFs coating sheets can reduce OTR even less than 1cc/m^2/day, which is considered suitable for packaging of most foods and medicines [57].

Typically, CNFs films are produced by aqueous dispersion and subsequent air drying of the sheets having high OTR (19.08cc/m2/day) that is not suitable for packaging. Two dimensional (2D) nanomaterials, comprising of graphene and MoS_2, have been synthesized to improve the oxygen barrier properties of the matrix film. These two dimensional structures increase the pathway distance for oxygen gas and reduce the flow. Graphene and MoS_2 are colored polymers having high coefficient for optical absorption. Colored polymers are cytotoxic and also decrease aesthetic value of packaging material [58]. The nanocellulose reinforcement of composite during preparation of films mainly for food packaging application is discussed below using some recent examples from literature.

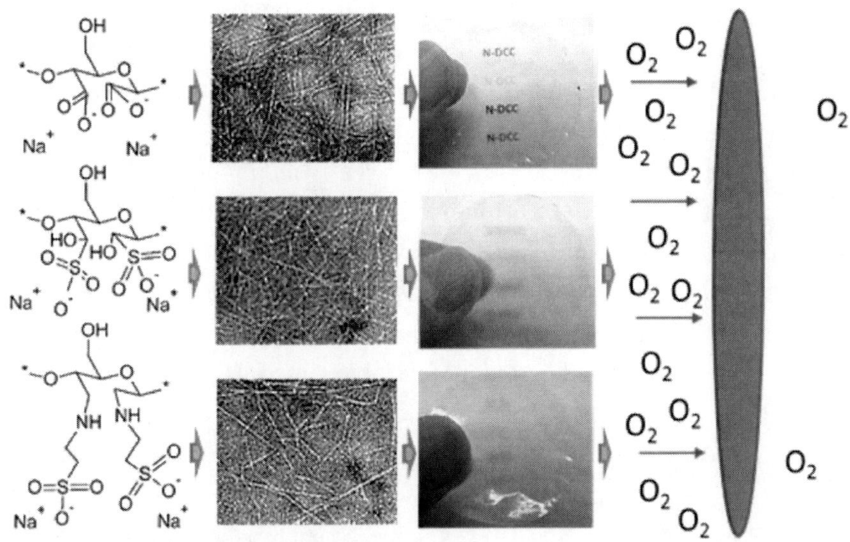

Figure 6. Oxygen barrier properties of nanocellulose films derived from pulp. Reproduced with permission from ACS [56].

Boron nitride nanosheet (BNNS) is a form of 2D nanomaterial having much better properties for packaging than conventional 2D nanomaterials. Its cost of production on commercial scale is much lower than graphene. Its color is white with lower coefficient of absorption and less cytotoxic than graphene. BNNS containing CNFs composite film can attain a low OTR of ~4.7cc/m^2/day. Nguyen et al. found that CNF/BNNS composite film had good oxygen barrier characteristics with OTR of <5cc/m^2/day which is considered suitable for packaging of meat and cheese. Addition of BNNS to CNFs not only improved the oxygen barrier properties but also increased the tensile strength of the material without sacrificing the elongation of fibers. Composite film also had no toxicity to MC3T3 cells showing great potential of film for food packaging [54].

Shimizu et al. conducted an experiment while soaking, 2,2,6,6-tetramethylpiperidine-1-oxyl (TEMPO)-oxidized cellulose nanofibril (TOCN) films with sodium carboxylate groups (TOCN–COONa), in aqueous solutions of $AlCl_3$, $MgCl_2$, $FeCl_3$ and $MgCl_2$ to change the ions to TOCNs–COOM films. Dry TOCN films associated with aluminium and calcium carboxylates exhibited very low oxygen permeability of 0.08 and 0.15mL/m^2/day/ kPa. Extremely low OTR in the film is attributed to the interfibrillar cross linkages with multivalent ions which make it suitable for packaging for numerous wet and humid conditions [59].

Xia et al. conducted an experiment to design cellulose nanofiber film with extremely low oxygen permeability and high transparency at high humidity level. The sheet was made from cellulose nanofibers with subsequent thermal treatment. Some reducing agents were used in sheet prior to thermal treatment to avoid discoloration during heating. This film had high transparency properties (85-90% at 600 nm). Standard electron microscope (SEM) images showed loss of protons during heat treatment resulted in high gas barrier and low water retention properties. The film had extremely low oxygen permeability (0.007 ml lm kPa^{-1} m^{-2} day^{-1}) at 23°C and 50% relative humidity (RH) by treating at 145°C. The film was found to have oxygen permeability of 0.584 mL kPa^{-1} m^{-2} day^{-1} at 80% relative humidity (RH), that is considered 100 times lower than plastic films i.e., polyvinyl chloride and polyethylene terephthalate [60].

Bardet et al. designed transparent films with cellulose nanofibers (CNF) which was strengthened with nanoclay (i.e., sodium montmorillonite, MMT). The neat film produced had good mechanical properties and excellent barrier characteristics for air at low moisture content level (<50%

RH). The film reduced the air barrier properties at high moisture contents, so it cannot be used as a packaging material at high moisture levels. MMT were used as nanofiller to retain the barrier qualities at high water contents. This film had increased density, and crystallinity that helped to improve the barrier qualities. The film had improved the transparency and mechanical properties to be used as packaging material for food products at high moisture contents level [61].

Yu et al. developed biopolymer-based nanocomposite film utilizing cellulose nanofibers (CNF), corn starch and chitosan. The film produced had high rigidity due to more hydrogen bonding in cellulose nanofibers (≥60%). High contents of CNFs in the film improved its barrier capacity against light, oxygen and humidity. The presence of 80-100% CNFs in the sheet improved the antimicrobial activity up to 2 log CFU/g on day 8 on beef model. Increased microbial activity may be attributed to improved oxygen barrier and more positive charges in the film. So CNFs can be used to improve the oxygen barrier property and antimicrobial effects in a biopolymer based nanocomposite film which makes it suitable for active packaging of food products [62].

Mondragon et al. conducted an experiment to assess the air and moisture permeability utilizing different techniques while adding nanocellulose fibers in gelatin matrix. Addition of nanocellulose to gelatin matrix significantly improved the oxygen barrier properties of gelatin matrix [63].

Wu et al. showed that 2,2,6,6-tetramethylpiperidine-1oxy radical (TEMPO) oxidized cellulose nanofibrils (TOCNs)/montmorillonite (MTM) composite films can be synthesized by simple mixing and drying process. Composite film exhibits excellent mechanical and oxygen barrier properties while having low densities [64].

Svagan et al. showed that nanocellulose composite capsules can be synthesized with liquid hexadecane core by direct mini emulsion polymerization. By mixing hexadecane cores with TEMPO-NFC suspension and then subsequent passing through the membrane resulted into formation of bigger capsules (8.3 ± 2.5 m) with higher nanocellulose contents. These films showed ultrahigh oxygen barrier properties at low moisture content level. It had <0.0001 cm^3 (STP) $mm/m^2/day/atm$ at 20% relative humidity

(RH) and 0.0054 cm^3 (STP) $mm/m^2/day/atm$ at 50% RH and 23°C. Having these properties, NFCs based capsules can be utilized for protection of oxygen sensitive molecules in dry state. These nanocellulose based capsules can be synthesized in modified ways to tune and increase the oxygen barrier properties of the capsule. These capsules can be used to increase the shelf-life of some sensitive molecules e.g., polyunsaturated oils (fish oil) in food and pharmaceutical industries [65].

Liimatainen et al. studied that NFCs talc hybrid films with lamellar brick and mortar structure were synthesized from cellulose nanofibers and talc platelets utilizing vacuum filtration method. Talc aggregates were disintegrated into well dispersed platelets to improve homogeneity of hybrid structure. These hybrids had talc contents 1-50% by weight, and showed flexibility in bending, tensile strength and Young's modulus values up to 211 MPa and 12 GPa respectively. Improvement of composite sheet material is attributed to the fine disintegration of individual talc platelets, their fine and even distribution and are tightly embedded on the NFCs network. This resulted into low pore size of the film and provide better oxygen barrier properties. [66].

Other Barrier Properties

Nanocellulose based composite films and coatings were discussed regarding in previous section regarding their vapor barrier and oxygen barrier characteristics. This section elucidates research efforts made to reduce limonene vapor permeance (aroma loss) [44], to minimize the loss of ascorbic acid [67] and to produce oil and grease resistant films [68]. For practical relevance, it is important to study the effects of nanofillers in other properties along with oxygen and vapor barrier. Therefore, efforts were made by Cherpinski et al. to determine aroma barrier in films composed of polyester/nanocellulose composite [44]. This was evaluated in terms of permeance to limonene vapor (standard method). Polyhydroalkanoate-coated nanopapers exhibited slightly lower aroma barrier because the polyester has affinity towards limonene uptake. The permanence to limonene uptake was found to be 1.7×10^{-11} kg/ m^2 s^{-1} Pa^{-1}. Though the film demonstrated greater vapor and oxygen barrier properties, but still further surface modification of the film is required to attain optimum balance in barrier properties of water vapor, oxygen and aroma.

Dong et al. evaluated loss of ascorbic acid content in cucumbers [67]. Oxalic acid solution was used to extract ascorbic acid from the sample and titration was performed to determine quantity of extracted (loss) ascorbic acid. It was demonstrated that the loss of ascorbic acid was inhibited by coating with nanocellulose and chitosan. The rate of loss of ascorbic acid obtained was lower than that of pure chitosan film. It was drastically reduced as compared to control sample (without coating) as the ascorbic acid contents were reduced to lowest level of 0.017 mg/100 g sample in 2 days with no coating. On application of composite coating, it took 12-14 days to reach the same level. Thus, the shelf life of cucumber was enhanced with the nanocellulose/chitosan composite coating.

Tyagi et al. studied oil and grease resistance of composite coatings prepared from nanocellulose, bentonite nanoclay, proteins as amphiphilic polyelectrolytes and alkyl ketene dimer (AKD) as a hydrophobic binder [68]. The idea behind using these constituents was to achieve enhanced orientation and dispersion of nanocellulose and avoid physical imperfections of nanofillers and brittleness (due to crystallinity) in the composite. The oil and gas resistance of the films were evaluated using Scotchban test following the standard procedure (Kit Test TAPPI T559). The composites demonstrated good resistance against oil and grease.

Packaging Applications of Nanocellulose Based Films or Coatings

The purpose of packaging materials can be broadly divided into three categories. These are containment, communication and protection. Containment refers to adequate mechanical strength and thermal characteristics to hold a variety of food products (cold, hot, dense and lightweight food). Communication serves the purpose of presenting nutritional facts and other important details to consumers. Protection has the ability of packaging material to withstand mechanical shocks during transportation and storage. Another aspect of protection to increase the shelf life of products is permeability. Proper packaging prevents oxygen and water vapor from entering and preserves aroma and flavour. With the advent of nanotechnology, there has been advances in food packaging. Our focus in this section will be on nanocellulose based composites and coatings for improving barrier characteristics, which have direct implications on the shelf-life of food.

There is growing consumer demand for environmentally friendly packaging material with enhanced quality and an extended shelf life. Successful development of packaging material with all these characteristics also help in eliminating addition of chemical preservatives in food. This will serve the purpose of offering natural food and eliminate excess wastage.

Many synthetic films made from Polyethylene (PE), Poly-vinyl chloride (PVC), Polyester (PET films) have excellent barrier performances and are successfully commercialized. In the recent years, owing to ecological requirements and demand for sustainability, efforts have been made to replace synthetic polymers with natural and bio-based polymers such as chitosan and pectin. Coatings made from chitosan have been commercialized to extend shelf life of fruits and vegetables such as tomato, strawberry, cucumber etc. But the major problem of these coatings is inability to protect the product against the flow of oxygen and moisture. It can have detrimental effect on product life. Thus, nanocellulose was incorporated with chitosan [67] to obtain barrier against oxygen and water vapor. Yu et al. developed edible nanocomposite film with corn starch and chitosan which showed a great potential for active packaging of food products [62]. Chaichi et al. prepared pectin reinforced cellulose nanocrystal based bionanocomposite films to overcome the limitation of poor barrier property and hydrophilic nature of pure pectin film [69].

Transparency is another important requirement for food packaging applications. The visual appearance provides better scope to consumers for selection of products. In lieu of obtaining transparent film with good thermomechanical characteristics as well as superior barrier properties, Trigol et al. obtained transparent nanocomposite films through addition of nanocellulose and nanoclay to Poly (lactic acid) [49]. Pereda et al. obtained polyelectrolyte films with chitosan and olive oil reinforced by nanocrystals of cellulose [70]. The combination of oil and nanocellulose offered high water vapor barrier characteristics. It was suggested that the transparency which was reduced by lipid addition was balanced by cellulose contents [70].

Nanocellulose can also be functionalized with chemical entities to obtain tunable barrier properties. Modification of CNFs to obtain acetylated and carboxymethylated CNFs have been demonstrated by Sharma et al. with excellent oxygen barrier characteristics of membrane at higher relative humidity [71]. Rafieian et al. reinforced gluten with nanocrystals of carboxylated cellulose and obtained promising results for applications where water vapor permeability is required [41]. Bideau et al. developed protective coating for paper board using poly(pyrrole) and nanocellulose for packaging applications[72]. The coating helped to reduce water sensitivity in paperboard. Edible coating was formulated with cellulose nanofibers to improve barrier performance of coated grapes [50]. The incorporation of nanocellulose in various natural polymer based composites and coatings have found to offer suitable gas barrier performances. Large scale production and achievement of cost-effectiveness would surely make the realization of these products on commercial platform in coming decades.

Conclusion and Outlook

With the increase of consumers awareness about sustainability and renewability, the global food packaging market from bio-based origin is expected to have a positive impact. Studies have shown that CNCs and CNFs based materials for food packaging applications have potential to replace plastic packaging from petroleum resources with superior oxygen and water barrier properties. The improved barrier properties of nanocellulose based packaging is due to the formation of dense network of nanocellulose during processing which greatly reduce the space for gaseous molecules transportation. Additionally, the better dispersion/adhesion between nanocellulose and polymer films or coating gives better barrier properties. In the current industrial set up, use of nanocellulose with other petroleum based polymer shows the best way to make polymers for food packaging applications. The challenge is to use nanocellulose with petro-derived counterparts while maintaining the renewability and biodegradability along with better oxygen and water vapor barrier properties. During transition to bio-based economy, a hybrid petroleum/biopolymers material may be practical as petroleum-based materials are still produced at low cost with better barrier properties. However, more intensive research is needed for better understanding of gas barrier mechanism, relationship between CNCs/CNFs structure, processing and their barrier properties. Also, further research may be directed to moisture control effects in nanocellulose derived films or coatings.

REFERENCES

[1] Wang, J., Gardner, D. J., Stark, N. M., Bousfield, D. W., Tajvidi, M., Cai, Z., 2017. Moisture and oxygen barrier properties of cellulose nanomaterial-based films. *ACS Sustainable Chemistry & Engineering* 6, 49-70.

[2] https://www.marketwatch.com/press-release/food-container-market-size-to-reach-usd-34699-billion-by-2025-grand-view-research-inc-2017-01-02.

[3] Petersen, K., Nielsen, P. V., Bertelsen, G., Lawther, M., Olsen, M. B., Nilsson, N. H., Mortensen, G., 1999. Potential of biobased materials for food packaging. *Trends in Food Science & Technology* 10, 52-68.

[4] Youssef, A. M., El-Sayed, S. M., 2018. Bionanocomposites materials for food packaging applications: Concepts and future outlook. *Carbohydrate Polymers* 193, 19-27.

[5] Nair, S. S., Zhu, J., Deng, Y., Ragauskas, A. J., 2018. High performance green barriers based on nanocellulose. *Sustainable Chemical Processes* 2, 23, 1-7.

[6] Chae, Y., An, Y.-J., 2018. Current research trends on plastic pollution and ecological impacts on the soil ecosystem: A review. *Environmental Pollution* 240, 387-395.

[7] Shaghaleh, H., Xu, X., Wang, S., 2018. Current progress in production of biopolymeric materials based on cellulose, cellulose nanofibers, and cellulose derivatives. *RSC Advances* 8, 825-842.

[8] Kulkarni, A. G., Vedamurthy, A. B., 2018. Cellulolytic Microorganisms: A Review, in: *Handbook of Research on Microbial Tools for Environmental Waste Management*, IGI Global, 34-47.

[9] Khalil, H. A., Tye, Y. Y., Leh, C. P., Saurabh, C., Ariffin, F., Fizree, H. M., Mohamed, A., Suriani, A., 2018. *Cellulose reinforced biodegradable polymer composite film for packaging applications, in: Bionanocomposites for packaging applications*, Springer, 49-69.

[10] Gicquel, E., Martin, C., Yanez, J. G., Bras, J., 2017. Cellulose nanocrystals as new bio-based coating layer for improving fiber-based mechanical and barrier properties. *Journal of Materials Science* 52, 3048-3061.

[11] Stelte, W., Sanadi, A. R., 2009. Preparation and characterization of cellulose nanofibers from two commercial hardwood and softwood pulps. *Industrial & Engineering Chemistry Research* 48, 11211-11219.

[12] Nair, S. S., Zhu, J., Deng, Y., Ragauskas, A. J., 2014. Hydrogels prepared from cross-linked nanofibrillated cellulose. *ACS Sustainable Chemistry & Engineering* 2, 772-780.

[13] Syverud, K., Stenius, P., 2009. Strength and barrier properties of MFC films. *Cellulose* 16, 75-85.

[14] Joshi, M., Adak, B., Butola, B., 2018. Polyurethane nanocomposite based gas barrier films, membranes and coatings: A review on synthesis, characterization and potential applications. *Progress in Materials Science*.

[15] Misra, M., Pandey, J. K., Mohanty, A., 2015. *Biocomposites: design and mechanical performance*, Woodhead Publishing.

[16] Hiltner, A., Liu, R., Hu, Y., Baer, E., 2005. Oxygen transport as a solid-state structure probe for polymeric materials: A review. J*ournal of Polymer Science Part B: Polymer Physics* 43, 1047-1063.

[17] Lagaron, J., Catalá, R., Gavara, R., 2004. Structural characteristics defining high barrier properties in polymeric materials. *Materials Science and Technology* 20, 1-7.

[18] Drozdov, A. D., Christiansen, J. D., Gupta, R., Shah, A., 2003. Model for anomalous moisture diffusion through a polymer–clay nanocomposite. *Journal of Polymer Science Part B: Polymer Physics* 41, 476-492.

[19] Silvestre, C., Duraccio, D., Cimmino, S., 2011. Food packaging based on polymer nanomaterials. *Progress in Polymer Science* 36, 1766-1782.

[20] Choudalakis, G., Gotsis, A., 2009. Permeability of polymer/clay nanocomposites: a review. *European Polymer Journal* 45, 967-984.

[21] Rhim, J.-W., Ng, P. K., 2007. Natural biopolymer-based nanocomposite films for packaging applications. *Critical Reviews in Food Science and Nutrition*, 47, 411-433.

[22] Tang, X., Kumar, P., Alavi, S., Sandeep, K., 2012. Recent advances in biopolymers and biopolymer-based nanocomposites for food packaging materials. *Critical Reviews in Food Science and Nutrition* 52, 426-442.

[23] Chatterjee, U., Butola, B., Joshi, M., 2016. Optimal designing of polyurethane-based nanocomposite system for aerostat envelope. *Journal of Applied Polymer Science* 133, 43529.

[24] Saxena, A., Elder, T. J., Kenvin, J., Ragauskas, A. J., 22010. High oxygen nanocomposite barrier films based on xylan and nanocrystalline cellulose. *Nano-Micro Letters* 2, 235-241.

[25] Guo, J., Catchmark, J. M., 2012. Surface area and porosity of acid hydrolyzed cellulose nanowhiskers and cellulose produced by Gluconacetobacter xylinus. *Carbohydrate Polymers* 87, 1026-1037.

[26] Nair, S. S., Zhu, J., Deng, Y. Ragauskas, A. J., 2014. Characterization of cellulose nanofibrillation by micro grinding, *Journal of Nanoparticle Research* 16, 2349.

[27] Belbekhouche, S., Bras, J., Siqueira, G., Chappey, C., Lebrun, L., Khelifi, B., Marais, S., Dufresne, A., 2011. Water sorption behavior and gas barrier properties of cellulose whiskers and microfibrils films. *Carbohydrate Polymers* 83, 1740-1748.

[28] Spence, K. L., Venditti, R. A., Rojas, O. J., Habibi, Y., Pawlak, J. J., 2010. The effect of chemical composition on microfibrillar cellulose films from wood pulps: water interactions and physical properties for packaging applications. *Cellulose* 17, 835-848.

[29] Wijmans, J. G., Baker, R. W., 1995. The solution-diffusion model: a review. *J Membrane Sci.* 107, 1-21.

[30] Ojijo, V., Ray, S. S., 2013. Processing strategies in bionanocomposites. *Progress in Polymer Science* 38, 1543-1589.

[31] Kar, K. K., Rana, S., Pandey, J., 2015. *Handbook of polymer nanocomposites. Processing, performance and application, Volume B: Carbon Nanotube Based Polymer Nanocomposites*, Springer Heidelberg New York Dordrecht, London.

[32] Park, H. J., Chinnan, M. S., 1995. Gas and water vapor barrier properties of edible films from protein and cellulosic materials. *Journal of food Engineering*, 25 (1995) 497-507.

[33] Azeredo, H. M., Rosa, M. F., Mattoso, L. H. C., 2017. Nanocellulose in bio-based food packaging applications. *Industrial Crops and Products* 97, 664-671.

[34] Yang, Q., Fukuzumi, H., Saito, T., Isogai, A., Zhang, L., 2011. Transparent cellulose films with high gas barrier properties fabricated from aqueous alkali/urea solutions. *Biomacromolecules* 12, 2766-2771.

[35] Yang, Q., Saito, T., Isogai, A., 2012. Facile fabrication of transparent cellulose films with high water repellency and gas barrier properties. *Cellulose* 19, 1913-1921.

[36] Willberg-Keyriläinen, P., Vartiainen, J., Pelto, J., Ropponen, J., 2017. Hydrophobization and smoothing of cellulose nanofibril films by cellulose ester coatings. *Carbohydrate Polymers* 170, 160-165.

[37] Hu, S., Gu, J., Jiang, F., Hsieh, Y.-L., 2016. Holistic rice straw nanocellulose and hemicelluloses/lignin composite films. *ACS Sustainable Chemistry & Engineering* 4, 728-737.

[38] Aulin, C., Ström, G. R., 2013. Multilayered alkyd resin/nanocellulose coatings for use in renewable packaging solutions with a high level of moisture resistance. *Industrial & Engineering Chemistry Research* 52, 2582-2589.

[39] Güner, F. S., Yağcı, Y., Erciyes, A. T., 2006. Polymers from triglyceride oils. *Progress in Polymer Science* 31, 633-670.

[40] Fogelstrom, L., Malmstrom, E., Johansson, M., Hult, A., 2010. Hard and flexible nanocomposite coatings using nanoclay-filled hyperbranched polymers. *ACS Applied Materials & Interfaces* 2, 1679-1684.

[41] Rafieian, F., Shahedi, M., Keramat, J., Simonsen, J., 2014. Mechanical, thermal and barrier properties of nano-biocomposite based on gluten and carboxylated cellulose nanocrystals. *Industrial Crops and Products* 53, 282-288.

[42] Khan, R. A., Salmieri, S., Dussault, D., Uribe-Calderon, J., Kamal, M. R., Safrany, A., Lacroix, M., 2010. Production and properties of nanocellulose-reinforced methylcellulose-based biodegradable films. *Journal of Agricultural and Food Chemistry* 58, 7878-7885.

[43] Pereira, P. H., Waldron, K. W., Wilson, D. R., Cunha, A. P., de Brito, E. S., Rodrigues, T. H., Rosa, M. F., Azeredo, H. M., 2017. Wheat straw hemicelluloses added with cellulose nanocrystals and citric acid. Effect on film physical properties. *Carbohydrate Polymers* 164, 317-324.

[44] Cherpinski, A., Torres-Giner, S., Vartiainen, J., Peresin, M. S., Lahtinen, P., Lagaron, J. M., 2018. Improving the water resistance of nanocellulose-based films with polyhydroxyalkanoates processed by the electrospinning coating technique. *Cellulose* 25, 1291-1307.

[45] Chi, K., Catchmark, J. M., 2018. Improved eco-friendly barrier materials based on crystalline nanocellulose/chitosan/carboxymethyl cellulose polyelectrolyte complexes. *Food Hydrocolloids* 80, 195-205.

[46] Saxena, A., Ragauskas, A. J., 2009. Water transmission barrier properties of biodegradable films based on cellulosic whiskers and xylan. *Carbohydrate Polymers* 78, 357-360.

[47] Qazanfarzadeh, Z., Kadivar, M., 2016. Properties of whey protein isolate nanocomposite films reinforced with nanocellulose isolated from oat husk. *International Journal of Biological Macromolecules* 91, 1134-1140.

[48] Sudhakaran Nair, S., Chen, H., Peng, Y., Huang, Y., Yan, N., 2018. Polylactic acid biocomposites reinforced with nanocellulose fibrils with high lignin content for improved mechanical, thermal and barrier properties. *ACS Sustainable Chemistry & Engineering*.

[49] Trifol, J., Plackett, D., Sillard, C., Szabo, P., Bras, J., Daugaard, A. E., 2016. Hybrid poly (lactic acid)/nanocellulose/nanoclay composites with synergistically enhanced barrier properties and improved thermomechanical resistance. *Polymer International* 65, 988-995.

[50] Silva-Vera, W., Zamorano-Riquelme, M., Rocco-Orellana, C., Vega-Viveros, R., Gimenez-Castillo, B., Silva-Weiss, A., Osorio-Lira, F., 2018. Study of Spray System Applications of Edible Coating Suspensions Based on Hydrocolloids Containing Cellulose Nanofibers on Grape Surface (Vitis vinifera L.). *Food and Bioprocess Technology,* 1-11.

[51] Mirmehdi, S., de Oliveira, M. L. C., Hein, P. R. G., Dias, M. V., Sarantópoulos, C. I. G. D. L., Tonoli, G. H. D., 2018. Spraying Cellulose Nanofibrils for Improvement of Tensile and Barrier Properties of Writing & Printing (W&P) Paper. *Journal of Wood Chemistry and Technology,* 1-13.

[52] Herrera, M. A., Mathew, A. P., Oksman, K., 2017. Barrier and mechanical properties of plasticized and cross-linked nanocellulose coatings for paper packaging applications. *Cellulose* 24, 3969-3980.

[53] Aulin, C., Karabulut, E., Tran, A., Wågberg, L., Lindström, T., 2013. Transparent nanocellulosic multilayer thin films on polylactic acid with tunable gas barrier properties. *ACS Applied Materials & Interfaces* 5, 7352-7359.

[54] Nguyen, H. L., Hanif, Z., Park, S.-A., Choi, B. G., Tran, T. H., Hwang, D. S., Park, J., Hwang, S. Y., Oh, D. X., 2018. Sustainable Boron nitride nanosheet-reinforced cellulose nanofiber composite film with oxygen barrier without the cost of color and cytotoxicity. *Polymers* 10, 501, 1-11.

[55] Xiao, X., Zeng, Z., Xiao, S., 2008. Behavior and products of mechanochemical dechlorination of polyvinyl chloride and poly (vinylidene chloride), *Journal of Hazardous Materials* 151, 118-124.

[56] Sirviö, J. A., Kolehmainen, A., Visanko, M., Liimatainen, H., Niinimäki, J., Hormi, O. E. O., 2014. Strong, self-standing oxygen barrier films from nanocelluloses modified with regioselective oxidative treatments. *ACS Applied Materials & Interfaces* 6, 14384-14390.

[57] Chen, Y. W., Lee, H. V., 2018. Revalorization of selected municipal solid wastes as new precursors of "green" nanocellulose via a novel one-pot isolation system: a source perspective. *International Journal of Biological Macromolecules* 107, 78-92.

[58] Chen, C., He, Y., Xiao, G., Xia, Y., Li, H., He, Z., 2018. Two-dimensional hybrid materials: MoS2-RGO nanocomposites enhanced the barrier properties of epoxy coating. *Applied Surface Science* 444, 511-521.

[59] Shimizu, M., Saito, T., Isogai, A., 2016. Water-resistant and high oxygen-barrier nanocellulose films with interfibrillar cross-linkages formed through multivalent metal ions. *Journal of Membrane Science* 500, 1-7.

[60] Xia, J., Zhang, Z., Liu, W., Li, V. C., Cao, Y., Zhang, W., Deng, Y., 2018. Highly transparent 100% cellulose nanofibril films with extremely high oxygen barriers in high relative humidity. *Cellulose* 25, 4057-4066.

[61] Bardet, R., Reverdy, C., Belgacem, N., Leirset, I., Syverud, K., Bardet, M., Bras, J., 2015. Substitution of nanoclay in high gas barrier films of cellulose nanofibrils with cellulose nanocrystals and thermal treatment. *Cellulose* 22 1227-1241.

[62] Yu, Z., Alsammarraie, F. K., Nayigiziki, F. X., Wang, W., Vardhanabhuti, B., Mustapha, A., Lin, M., 2017. Effect and mechanism of cellulose nanofibrils on the active functions of biopolymer-based nanocomposite films. *Food Research International* 99, 166-172.

[63] Mondragon, G., Peña-Rodriguez, C., González, A., Eceiza, A., Arbelaiz, A., 2015. Bionanocomposites based on gelatin matrix and nanocellulose. *European Polymer Journal* 62, 1-9.

[64] Wu, C.-N., Saito, T., Fujisawa, S., Fukuzumi, H., Isogai, A., 2012. Ultrastrong and high gas-barrier nanocellulose/clay-layered composites, *Biomacromolecules* 13, 1927-1932.

[65] Svagan, A. J., Koch, C. B., Hedenqvist, M., Nilsson, F., Glasser, G., Baluschev, S., Andersen, M. L., 2016. Liquid-core nanocellulose-shell capsules with tunable oxygen permeability. *Carbohydrate Polymers* 136, 292-299.

[66] Liimatainen, H., Ezekiel, N., Sliz, R., Ohenoja, K., Sirviö, J. A., Berglund, L., Hormi, O., Niinimäki, J., 2013. High-strength nanocellulose–talc hybrid barrier films. *ACS Applied Materials & Interfaces* 5, 13412-13418.

[67] Dong, F., Li, S., Jin, C., Liu, Z., Zhu, K., Zou, H., Wang, X., 2016. Effect of nanocellulose/chitosan composite coatings on cucumber quality and shelf life. *Toxicological & Environmental Chemistry* 98, 450-461.

[68] Tyagi, P., Hubbe, M. A., Lucia, L., Pal, L., 2018. High performance nanocellulose-based composite coatings for oil and grease resistance. *Cellulose* 25, 3377-3391.

[69] Chaichi, M., Hashemi, M., Badii, F., Mohammadi, A., 2017. Preparation and characterization of a novel bionanocomposite edible film based on pectin and crystalline nanocellulose. *Carbohydrate Polymers* 157, 167-175.

[70] Pereda, M., Dufresne, A., Aranguren, M. I., Marcovich, N. E., 2014. Polyelectrolyte films based on chitosan/olive oil and reinforced with cellulose nanocrystals. *Carbohydrate Polymers* 101, 1018-1026.

[71] Sharma, S., Zhang, X., Nair, S. S., Ragauskas, A., Zhu, J., Deng, Y., 2014. Thermally enhanced high performance cellulose nano fibril barrier membranes. *RSC Advances* 4, 45136-45142.

[72] Bideau, B., Loranger, E., Daneault, C., 2018. Nanocellulose-polypyrrole-coated paperboard for food packaging application. *Progress in Organic Coatings* 123 128-133.

In: Cellulose to Nanocelluloses
Editor: Tri-Dung Ngo
ISBN: 978-1-53615-057-5
© 2019 Nova Science Publishers, Inc.

Chapter 9

A Review on Mechanical Behavior of Cellulose and Nanocellulose Composites

***Dinh-Chi Pham*[1,*] *and Tri-Dung Ngo*[2]**
[1]Global Engineering and Material Inc., Princeton, NJ, US
[2]Biomass Conversion and Processing, InnoTech Alberta, Edmonton, Alberta, Canada

Abstract

There are accelerating interests in the use of cellulose fiber-reinforced polymer composite materials in the past decades due to their great abundance, affordable cost, and their light-weight, recyclable and biodegradable characteristics. The research on nanocellulose composites with polymers has aslo made significant contributions to the development of functional and sustainable materials. This chapter attempts to provide a short summary on the mechanical properties and behavior of cellulose fibers, nanocellulose, and the cellulose and nanocellulose composites.

* Corresponding Author Email: dcpham@gem-innovation.com.

Addtionally, the challenges associated with cellulose and nanocellulose based composites with and without surface modifications are discussed.

Keywords: cellulose, nanocellulose, composites, biocomposite, nanocomposite

INTRODUCTION

Cellulose fibers can be obtained from many sources such as plant-based materials, algae, bacteria, bamboo, cotton, fungi, tunicates and wood [1, 2]. They have wide applications in the forest products, paper and textile industries and are increasingly used as the reinforcements in composite materials. There has been a growing interest in natural fibers and cellulose composite materials in research and industrial applications because they are widely available, lightweight, cheap, renewable, recyclable and biodegradable. However, the lack of good interfacial adhesion, moisture absorption, quality variations, low thermal stability have limitted their wide applications. Pretreatments of the cellulose fibers can be performed to modify the fiber surface and thus enhance their surface roughness and properties [3].

Cellulose is a straight chain polymer made of repeated glucose molecules and each cellulose molecule is rod-like and long. Cellulose chains are linked by hygrogen bonds, and the hydrogen bond network is responsible for cellulose mechanical properties. The native form of cellulose consists of a triclinic unit cell with single chain (Iα crystal form) and a monoclinic unit cell with two chains (Iβ crystal form) [4]. Iα forms are found in most bacteria and algae cellulose while Iβ are mainly found on plant and animal cellulose. X-ray diffraction measurements and theoretical and simulation models have been used to measure the elastic modulus of cellulose. An example of the cellulose modulus of 140GPa was reported by Sakurada et al. [5] through X-ray diffraction which was later verified by molecular simulations reported in Tanaka et al. [6]. The combined lower density and high strength of crystalline cellulose provides celluloses with a high specific modulus that are comparable to engineering reinforcement materials. Bouligand structure was reported in [7] as the most microstructural natural material possessing high strength and toughness. Cellulose is also considered as a fascinating and inexhaustible biopolymer [8]. Recently, application of all-cellulose composites have gained increased attentions. By using similar cellulosic materials for both matrix and reinforcement, the fiber-matrix adhesion is shown to be improved than that for cellulose-reinforced polymer composites.

There still remains a gap for traditional cellulose materials to meet increasing demands for high performance materials used in industries. Currently, the functionality and high strength to weight ratio of cellulose materials are being exploited from their hierarchical structure designs spanning from nano to macro length scales. Cellulose nanofibers are extracted from the nanoscale, and their high stiffness can be tailored by ulitilizing of the high crystallinity of individuallized nanofibers while reduction of the morphous parts in its hierarchical structures [9]. In this chapter, we provide a summary on the mechanical properties of cellulose fibers and nanocellulose as compared to conventional fibers such as glass or carbon fibers. In addition, the mechanical properties of natural composites reinforced by cellulose fibers and nanocellulose are reviewed, highlighting the enhanced properties that cellulose nanoparticles are able to provide over the engineered fiber reinforced polymer composites such as glass fiber-reinforced polymer (GFRP) or carbon fiber-reinforced polymer (CFRP) composites. Lastly, some insights into the properties obtained by all-cellulose composites and their modifications are given, showing the potentials of such monolithic cellulose composites in bioindustry and textile industry.

CELLULOSE FIBER AND NANOFIBER

The properties of natural fibers are first summarized herein. It has been reported that the properties of cellulose fibers are affected by many factors such as climate, harvest, location, maturity, variety, decortications, disintegration, fiber modification, textile, and technical processes [3]. Among the natural fibers, flax fibers are relatively strong with its elementary tensile strength being in the order of 1500MPa [10]. The modulus of flax fibers is dependent on the fiber's diameter, ranging from 39GPa for 35μm fiber's diameter to 78GPa for 5μm fiber's diameter [11, 12]. Soft-wood-Kraft-fibers and flax were reported in [13] to have similar strength and stiffness as E-glass fibers. The elastic modulus of wood is about 10GPa. Through chemical-pulping process, cellulose fibers can be separated from woods with a modulus up to 40GPa. The microfibrils extracted from the cellulose can have an elastic modulus of 70GPa and the cellulose chain can be approximated to have a modulus of 250GPa. However, current technology to separate these from microfibrils is still under developing [14]. One of the main factors affect the tensile strength of natural fibers is the fiber length and reinforcing efficiency. Mieck et al. [15] and Mukherjee et al. [16] highlighted that flax fibers possess moderately high-specific strength and stiffness but the tensile strength of flax fiber is much dependent on the fiber length. In constrast, the pineapple fiber strength is less dependent. Table 1 summarizes the properties of engineering matertials as compared to cellulose fibers. It can be seen that crystalline cellulose has an elastic modulus comparable to Kevlar but higher than glass fibers and lower than carbon fibers.

Table 1. Properties of cellulose fibers and engineering materials

Material	Elastic modulus (GPa)	Density (mg/m^3)	Specific modulus ($GPa.mg^{-1}m^3$)	References
Aluminum	69	2.7	26	[17]
Steel	200	7.8	26	[17]
Glass fiber	69	2.5	28	[17]
E-Glass fiber	73	2.54	28.74	[18]
S-Glass fiber	10	2.49	4.02	[18]
Carbon fiber	235	1.81	129.83	[18]
Kevlar 49	131	1.45	90.34	[18]
Crystalline cellulose fiber	138	1.5	92	[17]

Table 2. Properties of cellulose nanofiber

Material	Axial modulus (GPa)	Transverse modulus (GPa)	Failure srength (MPa)	Failure strain (%)	References
Wood fiber (WF)	14 - 27		300 - 1400	4 - 23	[9, 19]
Plant fiber (PF)	5 - 45		300 - 80	1.3 - 8	[9]
Microcrystalline cellulose (MFC)	25 ± 4				[20]
Cellulose nanocrystals (CNC) Plant Wood (PW)	57, 105	18 - 50			[9, 21, 22]
Bacterial cellulose (BC)	78 ± 17 114				[23, 24]
Cellulose Iβ Experiment Simulation	120 - 138 220 ± 50 110 - 173 137 - 168	15 ± 1 10 - 50	6.5 - 7.7		[9, 25-27]
Cellulose Iα	128 - 155	5 - 8			[25, 26]
Cellulose II Experiment Simulation	9 - 90 98 - 109	17 - 31	4.9 - 5.4		[25, 28, 30]

Nanocellulose can be prepared from woodpulp and any cellulose source material. The term nanocellulose or cellulose nanoparticles (CN) refers to several of the particle types that have at least one dimension in the nanoscale. They include microfibrillated cellulose (MFC), nanofibrilated cellulose (NFC), cellulose nanocrystals (CNC), algae cellulose particles (AC) and bacterial cellulose particles (BC). Table 2 presents the properties of celllose nanomaterials in comparison to the micron and macrosized scaled particles such as wood fibers (WF) and plant fibers (PF). The material properties of nanocellulose materials are affected by the particle types, the crystal structure (Iα, Iβ, II) and the measurement techniques. The CNC's axial modulus is seen at least twice as much as that of the MFC. CNC has attractive mechanical properties to be used in material applications. Films made from CNC can have a high stiffness up to 100GPa. The elastic properties of nano cellulose materials are often measured by X-ray diffraction and X-ray scattering methods [9].

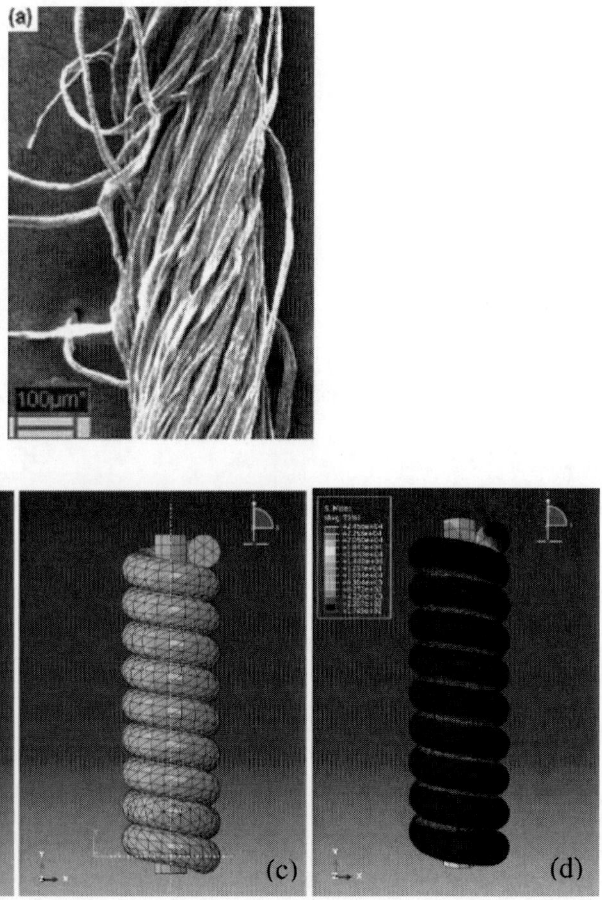

Figure 1. SEM image (x150) of bamboo showing uniform and compactly twisted fibers [34] (a) and examples of finite element analysis of bamboo threads with high twisting angle (b), low twising angle (c) and corresponding stress analysis (d) [35].

In addition to the experimental techniques for characterization of cellulose mechanical properties, analytical models and simulation models can be used to analyze the behavior and the effective properties of cellulose materials. Theoretical models using the rule of mixture or Halpin-Tsai model [31] have been employed to estimate the composite properties. On the other hand, simulation models using molecular dynamics to model atomic and molecular motions [6] or continuum mechanic approaches have also been adopted. The continuum mechanics models utilize a representative volume element (RVE) of the composite material to model the constituents' behavior and their interfaces [32, 33]. Finite element model of individual fibers and matrix can be built and stress analysis can be performed to assess the behavior of the overall composite material. Figure 1 illustrates an example of finite element analysis of bamboo fibers in Abaqus, evalulating the optimum fiber twisting angles and the number of threads to achieve the desired fiber strength. The fibers are moded with 3D solid elements consisting of one straight fiber and a number of threads helically wound around the longitudinal axis with a pre-defined helix angle. The fibers are assigned with Bamboo material, and contact algorithm is defined between individual fibers. Failure analyses of the fiber bundle subject to tension are carried out to predict its longitudinal strength. Parametric studies and optimization with respect to the fiber properties, the number of threads and their corresponsing twisting angles can then be performed and automated with Python scripts following the method presented by Pham et al. [35] to achieve the best bamboo configurations meeting the design requirements.

Cellulose Fiber Reinforced Composites

When considering the advantage of their low specific mass, natural composites are competitive to glass fiber-reinforced plastics (GFRP) in some applications. Bledzki and Gassan [36] demonstrated the tensile strength, flexural modulus and impact resistance of natural-fiber composites to be comparable to GFRP. Wambua et al. [37] studied the effect of the volume fraction of the constituents in natural fiber composites and reported an increase in the tensile strength of natural composites with increasing fiber volume fraction. The hemp/polypropylene (PP) composite with a 30% fiber volume fraction was found to exhibit a similar tensile strength of 52MPa as glass-reinforced composites with the same volume fraction. In addition, hemp and kenaf-polypropylene composites displayed a high tensile modulus of 6.8GPa as compared to 6.2GPa of equivalent GFRP. Bos et al. [38] revealed an increase in the modulus and adhesion of flax/PP composites with maleic-anhydride grafted polypropylene. Garkhail et al. [39] investigated the effect of fiber length on the composites strength of flax-fiber-reinforced PP composites and found the stiffness of a flax/PP composite to be comparable to E-glass-based composites. Nishino [40] analyzed the mechanical properties of kenaf/poly-L-lactide (PLLA) composites and suggested an increase in composite modulus with an increase of volume action till a threshold after which further increase of the volume fraction leads to an adverse effect on composite properties. Espert et al. [41] studied the water content effect on cellulose/PP composites. More significant effects were found for the Young's modulus of the cellulose PP than the tensile strength. Thwe and Liao [42] investigated the aging effect on bamboo-fiber composites and observed that the mechanical properties of bamboo composites decayed upon aging time and temperature. A coupling agent in the matrix material, e.g., maleic anhydride polypropylene, was seen to enhance the composite strength and stiffness.

Apart from mechanical static tests, there are limited studies on the dynamic behavior such as the impact behavior of natural-fiber reinforced composites. Wambua et al. [37] reviewed the impact performance of kenaf-, coir-, sisal-, hemp-, and jute reinforced polypropylene. The results showed that natural fiber composites provide relatively low impact strengths as GFRP. Pavithran et al. [43] measured the fracture energies for sisal, pineapple, banana, and coconut fiber-polyester composites in a Charpy impact test. It was revealed that fibers with higher fibril angles can produce higher fracture-toughness than those with small spiral angle. Tobias [44] later analyzed the effects of fiber content, length and angle on the impact strength of banana-fiber composites. It was concluded that smaller fiber lengths and higher fiber volume resulted in greater impact strength. In contrast, Garkhail et al. [39] suggested that the impact strength of flax/PP composites increases with increasing fiber length until a certain level after which the impact performance drops depending on the fiber pretreatment and the fiber/matrix interface adhesion. Mueller [45] investigated the thermal effect on the impact strength of hemp-, flax- and kenaf-polypropylene composites by compression molding. An optimum temperature was observed for each material to achieve its best impact strength, and the higher fineness fiber resulted in higher impact performance. De Bruijn [46] reviewed the effect of water absorption and temperature on the impact responses of natural-fiber thermoplastics. The impact strength of natural fiber composites was found about 20-25% of equivalent glass-reinforced thermoplastic. It was also shown that using a coupling agent may increase the composite interface strength and impact strength.

Compared to the studies on impact of natural composites, fatigue behavior is much less in the literature. Gassan [47] investigated the fatigue behavior of flax and jute epoxy composites and found it to be relied on the fiber type, volume fraction, fiber properties and interface properties. Natural fiber-reinforced plastics with stronger fiber mechanical properties, high fiber-matrix interface strength and or higher fiber weight fractions can sustain higher failure. More specifically, unidirectional natural fiber composites were less sensitive to fatigue-induced damage than natural woven fabrics. Savastano et al. [48] conducted experimental studies of fatigue crack growth in cementitious matrices reinforced with banana, sisal and bleached eucalyptus pulp. Three stages of fatigue crack growth were investigated: an initially decelerated growth, a steady crack growth, and a catastrophic crack growth. The fatigue life of sisal and banana fiber composite was found in the steady-state crack growth. Similarly to GFRP and CFRP, the fatigue crack growth in natural composites was dominated by matrix cracking, crack deflection around fibers, and crack-bridging between the fibers. Towo and Ansell [49] analyzed the effect of surface modification on the fatigue performance of sisal/epoxy composite. They showed that NaOH surface treatment may enhance the material tensile modulus but not improve the fatigue life of the sisal fiber composites. Sisal fiber composite was found to behave similarly as conventional synthetic fiber composites. Besides, the fatigue properties of flax/polyester was reported in [49] with alkali-treated and untreated fibers. The fiber treatment was observed to have a profound effect on the fatigue life of the flax/polyester materials. It was also highlighted that the polyester matrix coupons had lower life than the epoxy coupons. Yuanjian and Isaac [50] then compared the fatigue behavior between hemp- and flax-reinforced polyester composites. A steeper slope of the S-N curve was found for the hemp-fiber composite, indicating its faster strength degradation than the flax composite under cyclic loadings. However, the fatigue performance of the hemp composite was comparable and slightly greater than those of the glass fiber composite.

The tensile behavior of the jute fiber reinforced epoxy composite was experimentally investigated and numerically studied by finite element analyses in Ahmad et al. [51]. The effects of jute volume fractions, ply orientations, jute reinforcement types (unidirectional, woven mat, short fibers) and weft fiber angle were analyzed. By assuming the interface perfect bonding, it was found that the tensile strength of the jute woven mat composites is greater than that of the unidirectional jute composites. In addition, the laminate strength will increase with increasing jute volume fraction and weft fiber angle up to 60 degrees. Lee et al. [52] further investigated the effects of interphase elastic modulus on the performance of cellulose fibers reinforced polypropylene (PP) matrix composite modified by maleic anhydride-grafted PP through nano-indentation tests and finite element analyses.

In general, natural or cellulose fiber composites are viable replacements for synthetic fiber composites due to their low specific mass and high stiffness. Tensile and impact tests of natural fiber composites have been performed under laboratory conditions but there was large scattering of the tested properties. More tests are essential to investigate the influences of the yarns/resin interface on the measurement of composite properties. Besides, there has been still limited research on the moisture effects and long-term durability of natural fiber composites under cyclic loadings. Further works are, therefore, intended to characterize the environmental effects on static behavior and long-term performance of natural fiber composites.

CELLULOSE NANOFIBER REINFORCED NANOCOMPOSITES

The use of nanoparticles reinforcements has proven to impart improved mechanical and thermal properties for nanocomposites. The effects of inclusion of nano particles in polymner result in an increase in the interfacial area, aspect ratio and their extent of dispersion and percolation [53]. Nanomaterials can include carbon nanotubes, glass fibers, exfoliated clay and cellulose nanocrystals [54]. One wide application of nanomaterials is barrier film where the nano particles provide improved barrier properties such as durability, robustness and weight. One of the widely used filler nanomaterial is exfoliated clay. It was reported in [55] that by inclusion of less than 5% clay in polyester, polyimide, polypropylene, an enhancement on thermal resistance and stiffness were achieved for films. Cellulose nanocomposites has been rapidly expanding with many new materials that incorporating nanosized cellulose as reinforcements in polymer matrices [56]. Chakrabarty et al. [57] reported that CN served as the crosslinker in the water-borne polyurethane and CN nanocomposite, showing about 86% and 100% improvement in the tensile strength and Young's modulus, respectively. The tensile strength of the composite improved by 70-80% with the addition of cellulose nano whiskers as compared with neat PP [58]. However, in order to fabricate well-defined nanocomposites with the nanocellulose particles, the selection of the polymer matrix, the type of nanoccellulose and their chemistry are always crucial. The factors will affect the polymer-filler compatibility and thus the desired reinforcement effect for specific applications.

The main challenge for nanoparticles is associated with their dispersion within a polymeric matrix. The mechanical behavior of cellulose nanofibers composites is dependent on the properties and the volume fraction of each component, their spatial arrangements, the fiber/matrix interface properties, and the process methods. Utilization of cellulose nanofibers (CNs) can be done via (1) adding CNs to polymeric matrix and compare the filled polymer behavior to the unfilled neat polymer; (2) using 100% cellulose-based materials for both fiber and matrix (or neat CN films) and (3) using the neat CN films with modification. The modified CN film is distinguished from the CN reinforced polymer composites due to their high CN volume fraction. The properties of CN composites are determined by the CN network formation.

Table 3. Summary of properties for CN reinforced polymer composites$^{\alpha}$, neat CN film$^{\beta}$, and modified neat CN film$^{\gamma}$ [9]

Particle Type	Elastic modulus (GPa)	Failure strength (MPa)	Failure strain (%)	Coefficient of thermal expansion
MCC$^{\alpha}$	1.2 - 10	20 - 202	2.7 - 180	
MFC$^{\alpha}$	0.093 - 10	1.7 - 113	1 - 400	27 - 31
MFC$^{\beta}$	1 - 17.5	30 - 155	2.5 - 11.5	15 - 25
MFC$^{\gamma}$	5 - 19	60 - 200	1.6 - 12	10 - 14
NFC$^{\alpha}$	0.003 - 6.2	5 - 92	1.7 - 970	12 - 28.5
NFC$^{\beta}$	6 - 15	95 - 240	2 - 10	2.7 - 22
NFC$^{\gamma}$	4 - 17	80 - 290	0.8 - 11	10 - 23
CNC$^{\alpha}$	0.0007 - 3.9	1.5 - 218	1.9 - 986	
CNC$^{\beta}$	6			
t-CNC$^{\alpha}$	0.0005 - 6	0.2 - 50	0.5 - 285	
t-CNC$^{\beta}$	5 - 10	40 - 70	0.5 - 5	
t-CNC$^{\gamma}$	0.5 - 11	16 - 160	1.5 - 7.7	
BC$^{\alpha}$	0.355 - 11	21 - 188	3.5 - 30	4 - 38
BC$^{\beta}$	10 - 35	87 - 510	1.1 - 4.4	2 - 3
BC$^{\gamma}$	5 - 25	300 - 510	3 - 4	8 - 10

A comparison on the mechanical properties of different CN composite types is summarized in Table 3. Those composites reinforced with nanofibers are denoted with suffix α while all-cellulose composites are with suffix β and the modified neat film is with γ. As can be seen in Table 3, the bacterial cellulose composites (BC$^{\alpha}$), which contain microfibrils with large length-to-diameter ratios, have a relatively low tensile strain. Microcrystalline cellulose composites (MCC$^{\alpha}$), which comprises bundles of multi-sized microfibrils with high cellulose content and crystallinity, display a similar elastic modulus but higher failure strain than BC$^{\alpha}$. Different from the MCC$^{\alpha}$, the microfibrillated cellulose composites (MFC$^{\alpha}$), which comprise of 100% cellulose fibers ranging from 10 - 100nm in width and 0.5 - 10 μm in length, possess a higher failure strain than the MCC$^{\alpha}$ composites while the differences on the axial modulus and tensile strengths between MFC$^{\alpha}$ and MCC$^{\alpha}$ composites are not clear. By using nanoparticles with finer diameters than MFC$^{\alpha}$, nanofibrilated cellulose composite (NFC$^{\alpha}$) can produce even higher failure strain. Cellulose nanocrystal composites (CNC$^{\alpha}$), which contains nanocrystaline celluloses about 3 - 5nm wide and 50 - 500nm long, can give much higher failure strains than any other nanocomposites (MCC$^{\alpha}$, MFC$^{\alpha}$, NFC$^{\alpha}$).

With both the fiber and matrix are made of cellulose materials, the neat CN films such as MFC$^{\beta}$, NFC$^{\beta}$, CNC$^{\beta}$, t-CNC$^{\beta}$ exhibit higher mechanical properties than the nanoparticle reinforced composites of the same type (i.e., MFC$^{\alpha}$, NFC$^{\alpha}$, CNC$^{\alpha}$, t-CNC$^{\alpha}$). Furthermore, the mechanical properties of the composite are found to increase with increasing degree of fibrillation (from WF to MFC to NFC). However, excessive mechanical processing of all-cellulose composites can reduce their properties due to the remnant porosity from the gaps within the cellulose particle network.

The all-cellulose composites can be modified by filling the remnant porosity with a given chemical such as NaOH or melamine formaldehyde, acrylic, epoxy, phenol-formaldehyde [9]. Most of the studies on modified cellulose films have been focused on films with more than 70% CN volume fraction that can help retain the CN network and the CN-CN interaction. Although the modified cellulose composites (MFC$^{\gamma}$, NFC$^{\gamma}$, CNC$^{\gamma}$, t-CNC$^{\gamma}$) generally show higher axial stiffness and failure strength than the cellulose nanofiber-reinforced composites (MFC$^{\alpha}$, NFC$^{\alpha}$, CNC$^{\alpha}$, t-CNC$^{\alpha}$) and the neat film (MFC$^{\beta}$, NFC$^{\beta}$, CNC$^{\beta}$, t-CNC$^{\beta}$) of the same type, they have not been fully exploited. It is also noted that the modification of CN film properties is highly relied on the specific chemical treatment and the process incorporating the chemical into the CN network.

CONCLUSION

Cellulose and nanocellulose-based materials have been found in various applications such as packaging products, automotive, construction and biomedical applications. When used as reinforcement in polymer composite materials, cellulose and nanocellulose materials enhance the stiffness and strength of the composite. Current technologies are towards utilizing cellulose-based materials with additional pre-treatment or modification, thus generating more new materials with desired properties. The main challenges have been seen to be associated with cellulose-based composites including (1) interface de-cohesion characterization and mechanics between the reinforced fibers and the surrounding matrix, (2) characterization of CNs dispersion within the matrix polymer, and (3) the role of hierarchical structure on resulting macroscopic properties. The mechanical properties of the CN composites are also found sensitive to the fiber arrangements, orientations, constituent's weight fractions, and the processing methods such as polymer solution casting. There are scatter on the measured properties of cellulose-based composites reported in the literature. When these properties are used for industry applications, reliability analyses on the material properties are needed for design of composites. Moreover, more characterization tests are deemed to characterize the interfaces between CN-CN and CN-matrix. With the emerging high-performance computing technologies, simulation models using molecular dynamic simulations and continuum damage mechanics can be applied to reduce testing efforts and provide more understanding on the cellulose structures and properties.

REFERENCES

[1] Nabi, S. D. & Jog, P. J. (1999). Natural fiber polymer composites: A review. *Advances in Polymer Technology*, *18*(4), 351-363.

[2] Georgopoulos, S. T., Tarantili, P. A., Avgerinos, E., Andreopoulos, A. G. & Koukios, E. G. (2005). Thermoplastic polymers reinforced with

fibrous agricultural residues. *Polymer Degradation and Stability*, *90*(2), 303-312.

[3] Kalia, S., Dufresne, A., Cherian, B. M., Kaith, B. S., Luc, A., et al., (2011). Cellulose-based bio- and nanocomposites: A review. *International Journal of Polymer Science*, Volume *2011*, Article ID 837875. Doi:10.1155/2011/837875.

[4] Andresen, M., Stenstad, P., Møretrø, T., Langsrud, S., Syverud, K., Johansson, L.-S., et al., (2007). Nonleaching antimicrobial films prepared from surface-modified microfibrillated cellulose. *Biomacromolecules*, *8*(7), 2149-2155.

[5] Ichiro, S., Yasuhiko, N. & Taisuke, I. (1962). Experimental determination of the elastic modulus of crystalline regions in oriented polymers. *Journal of Polymer Science*, *57*(165), 651-660.

[6] Tanaka, F. & Iwata, T. (2006). Estimation of the elastic modulus of cellulose crystal by molecular mechanics simulation. *Cellulose*, *13*, 509-517.

[7] Roman, M. (2009). Model Cellulosic Surfaces, ed., *American Chemical Society*, ISBN13: 9780841269651, DOI: 10.1021/bk-2009-1019.

[8] Lucia, A. L. & Rojas, O. J. (2009*). The Nanoscience and technology of Renewable Biomaterials*, ed., John Wiley & Sons, Ltd, Singapore.

[9] Moon, R. J., Martini, A., Nairn, J., Simonsen, J. & Youngblood, J. (2011). Cellulose nanomaterials review: structure, properties and nanocomposites. *Chemical Society Reviews*, *40*(7), 3941-3994.

[10] Bos, H. L., van den Oever, M. J. A. & Peters, O. C. J. J. (2002). Tensile and compressive properties of flax fibres for natural fibre reinforced composites. *Journal of Materials Science*, *37*(8), 1683-1692.

[11] Baley, C. (2002). Analysis of the flax fibres tensile behaviour and analysis of the tensile stiffness increase. *Composites Part A: Applied Science and Manufacturing*, *33*(7), 939-948.

[12] Andersons, J., Spārniņš, E. & Joffe, R. (2006). Stiffness and strength of flax fiber/polymer matrix composites. *Polymer Composites*, *27*(2), 221-229.

[13] Gassan, J. & Bledzki, A. K. (1996). Einfluß von haftvermittlern auf das feuchteverhalten naturfaserverst¨arkter kunststoffe. [Influence of adhesion promoters on the moisture behavior of natural fiber-reinforced plastics.] *Die Angewandte Makromolekulare Chemie*, *236*, 129-138.

[14] Maloney, T. M. (1995). in *International encyclopedia of composites*, S. M. Lee S. M., Rowell, R. M., Eds., p. 656, VCH Publishers, New York, USA, 1995.

[15] Mieck, K. P., Nechwatal, A. & Knobelsdorf, C. (1994). Potential applications of natural fibres in composite materials. *Melliand Textilberichte*, *75* (11), 892-898.

[16] Mukherjee, P. S. & Satyanarayana, K. G. (1986). An empirical evaluation of structure-property relationships in natural fibres and their fracture behaviour. *Journal of Materials Science*, *21*(12), 4162-4168.

[17] Ashby, M. F. & Jone, D. R. H. (1989). *Engineering materials 1: an introduction to their properties and applications*. Pergamon Press, Oxford.

[18] Daniel, I. M. (2006). *Engineering mechanics of composite materials/ Isaac M. Daniel*, Ori Ishai. New York: Oxford University Press 2nd Ed.

[19] Mott, L., Groom, L. & Shaler, S. (2002). Mechanical properties of individual southern pine fibers. Part II. Comparison of earlywood and latewood fibers with respect to tree height and juvenility. *Wood Fiber Sci.*, *34*, 221-237.

[20] Eichhorn, S. J. & Young, R. J. (2001). The Young's modulus of a microcrystalline cellulose. *Cellulose*, *8*(3), 197-207.

[21] Rusli, R. & Eichhorn, S. J. (2008). Determination of the stiffness of cellulose nanowhiskers and the fiber-matrix interface in a nanocomposite using Raman spectroscopy. *Applied Physics Letters*, *93*(3), 033111.

[22] Lahiji, R. R., Xu, X., Reifenberger, R., Raman, A., Rudie, A. & Moon, R. J. (2010). Atomic force microscopy characterization of cellulose nanocrystals. *Langmuir: the ACS Journal of Surfaces and Colloids*, *26*(6), 4480-4488.

[23] Guhados, G., Wan, W. & Hutter, J. L. (2005). Measurement of the elastic modulus of single bacterial cellulose fibers using atomic force microscopy. *Langmuir: the ACS journal of surfaces and colloids*, *21*(14), 6642-6.

[24] Hsieh, Y. C., Yano, H., Nogi, M. & Eichhorn, S. J. (2008). An estimation of the Young's modulus of bacterial cellulose filaments. *Cellulose*, *15*(4), 507-513.

[25] Eichhorn, S. J. & Davies, G. R. (2006). Modelling the crystalline deformation of native and regenerated cellulose. *Cellulose*, *13*(3), 291-307.

[26] Stephan, R. & Jürgen, B. (1995). Theoretical investigations on the structure and physical properties of cellulose. *Macromolecular Theory and Simulations*, *4*(4), 725-743.

[27] Diddens, I., Murphy, B., Krisch, M. & Müller, M. (2008). Anisotropic elastic properties of cellulose measured using inelastic X-ray scattering. *Macromolecules*, *41*(24), 9755-9759.

[28] Eichhorn, S. J., Young, R. J. & Davies, G. R. (2005). Modeling crystal and molecular deformation in regenerated cellulose fibers. *Biomacromolecules*, *6*(1), 507-513.

[29] Eichhorn, S. J., Sirichaisit, J. & Young, R. J. (2001). Deformation mechanisms in cellulose fibres, paper and wood. *Journal of Materials Science*, *36*(13), 3129-3135.

[30] Northolt, M. G., den Decker, P., Picken, S. J., Baltussen, J. J. M. & Schlatmann, R., (2005). *Polymeric and Inorganic Fibers*, vol. *178*, 1-108.

[31] Halpin, J. C. & Kardos, J. L. (1976). The Halping-Tsai equations a review. *Polymer Engineering and Science*, *16*(5), 344-352.

[32] Pham, D. C., Sun, X. S., Tan, V. B. C., Chen, B. & Tay, T. E. (2012). Progressive failure analysis of scaled double-notched carbon/epoxy composite laminates. *International Journal of Damage Mechanics*, *21*(8), 1154-1185.

[33] Pham, D. C. & Sun, X. S. (2012). Experimental and computational studies on progressive failure analysis of notched cross-ply CFRP

composite. *International Journal of Computational Materials Science and Engineering*, *1*(3), 1-15.

[34] Majumdar, A., Mukhopadhyay, S., Yadav R. & Mondal, A. K. (2011). Properties of ring-spun yarns made from cotton and regenerated bamboo fibres. *Indian Journal of Fibre & Textile Research*, *36*, 18-23.

[35] Pham, D. C. & Narayanaswamy, S. (2014). An optimization tool for impact analysis of composite structures. *Procedia Engineering*, *75*, 3-8.

[36] Gauthier, R., Joly, C., Coupas, A. C., Gauthier, H. & Escoubes, M. (1998). Interfaces in polyolefin/cellulosic fiber composites: Chemical coupling, morphology, correlation with adhesion and aging in moisture. *Polymer Composites*, *19*(3), 287-300.

[37] Wambua, P., Ivens, J. & Verpoest, I. (2003). Natural fibres: can they replace glass in fibre reinforced plastics? *Composites Science and Technology*, *63*(9), 1259-1264.

[38] Bos, H. L., Müssig, J. & van den Oever M. J. A. (2006). Mechanical properties of short-flax-fibre reinforced compounds. *Composites Part A: Applied Science and Manufacturing*, *37*(10), 1591-1604.

[39] Garkhail, S. K., Heijenrath, R. W. H. & Peijs, T. (2000). Mechanical properties of natural-fibre-mat- reinforced thermoplastics based on flax fibres and polypropylene. *Applied Composite Materials*, *7*(5), 351-372.

[40] Nishino, T. (2004). Natural fiber sources. In *Green composites: Polymer composites and the environment*, CRC Press, Boca Raton, Fla, USA.

[41] Espert, A., Vilaplana, F. & Karlsson, S. (2004). Comparison of water absorption in natural cellulosic fibres from wood and one-year crops in polypropylene composites and its influence on their mechanical properties. *Composites Part A: Applied Science and Manufacturing*, *35*(11), 1267-1276.

[42] Thwe, M. M. & Liao, K. (2003). Durability of bamboo-glass fiber reinforced polymer matrix hybrid composites. *Composites Science and Technology*, *63*(3), 375-387.

[43] Pavithran, C., Mukherjee, P. S. & Brahmakumar, M. (1991). Coir-glass intermingled fibre hybrid composites. *Journal of Reinforced Plastics and Composites*, *10*(1), 91-101.

[44] Tobias, B. C. (1993). *Tensile and Impact behaviour of natural fiber-reinforced composite materials. in Proceedings of the International Conference on Advanced Composite Materials*, Chandra, T.D.A. K. Ed., The Minerals, Metals and Materials Society.

[45] Mueller, H. D. (2004). *Improving the impact strength of natural fiber reinforced composites by specifically designed material and process parameters.* IJN Winter.

[46] de Bruijn, J. C. M. (2000). Natural fibre mat thermoplastic products from a processor's point of view. *Applied Composite Materials*, *7*(5), 415-420.

[47] Gassan, J. (2002). A study of fibre and interface parameters affecting the fatigue behaviour of natural fibre composites. *Composites Part A: Applied Science and Manufacturing*, *33*(3), 369-374.

[48] Savastano, H., Santos, S. F., Radonjic, M. & Soboyejo, W. O. Fracture and fatigue of natural fiber-reinforced cementitious composites. *Cement and Concrete Composites*, *31*(4), 232-243.

[49] Towo, A. N. & Ansell, M. P. (2008). Fatigue of sisal fibre reinforced composites: Constant-life diagrams and hysteresis loop capture. *Composites Science and Technology*, *68*(3), 915-24.

[50] Yuanjian, T. & Isaac, D. H. (2007). Impact and fatigue behaviour of hemp fibre composites. *Composites Science and Technology*, *67*(15), 3300-3307.

[51] Ahmad, F. & Bajpai, P. K. (2018). Evaluation of stiffness in a cellulose fiber reinforced epoxy laminates for structural applications: Experimental and finite element analysis. *Defence Technology*, *14*, 278-286.

[52] Lee, S. H., Wang, S., Pharr, G. M. & Xu, H. (2007). Evaluation of interphase properties in a cellulose fiber-reinforced polypropylene composite by nanoindentation and finite element analysis. *Composites Part A*, *38*, 1517-1524.

[53] Helbert, W., Cavaille, J. Y. & Dusfesne, A. (1996). Thermoplastic nanocomposites filled with wheat straw cellulose whiskers. Part I: Processing and mechanical behavior. *Polymer Composites*, *17*(4), 604-611.

[54] Sturcova, A., Davies, G. R. & Eichhorn, S. J. (2005). Elastic modulus and stress-transfer properties of tunicate cellulose whiskers. *Biomacromolecules*, *6*(2), 1055-1061.

[55] Elazzouzi-Hafraoui, S., Nishiyama, Y., Putaux, J. L., Heux, L., Dubreuil, F. & Rochas, C. (2008). The shape and size distribution of crystalline nanoparticles prepared by acid hydrolysis of native cellulose. *Biomacromolecules*, *9*(1), 57-65.

[56] Matos Ruiz, M., Cavaillé, J. Y., Dufresne, A., Gérard, J. F. & Graillat, C. (2000). Processing and characterization of new thermoset nanocomposites based on cellulose whiskers. *Composite Interfaces*, 7(2), 117-131.

[57] Chakrabarty, A. & Teramoto, Y. (2018). Recent advances in nanocellulose composites with polymers: A guide for choosing partners and how to incorporate them. *Polymers*, *10*, 517, doi:10.3390/polym10050517.

[58] Bahar, E., Ucar, N., Onen, A., Wang, Y., Oksüz, M., Ayaz, O., Ucar, M. & Demir, A. (2012). Thermal and mechanical properties of polypropylene nanocomposite materials reinforced with cellulose nano whiskers. *Journal of Applied Polymer Science*, *125* (4), 2882-2889. doi: 10.1002/app.36445.

In: Cellulose to Nanocelluloses
Editor: Tri-Dung Ngo
ISBN: 978-1-53615-057-5
© 2019 Nova Science Publishers, Inc.

Chapter 10

STABILITY AND MOBILITY OF FOAM GENERATED BY GAS-SOLVENT/NANOCELLULOSE UNDER RESERVOIR CONDITIONS

Chao Wang and Huazhou Andy Li*
Department of Civil and Environmental Enginering,
University of Alberta, Edmonton, Alberta

ABSTRACT

Foam can be a feasible means to remedy the low sweep efficiency of solvent flooding (e.g., CO_2 and/or C_3H_8) resulting from viscous fingering and gravity segregation. It works as a good mobility control agent that decreases the mobility difference between the displaced oil and the displacing agent. Solvent-alternating-surfactant is a feasible way to create foam under reservoir conditions. However, most of the surfactant-stabilized foams are normally unstable and thermally degradable, especially with the presence of salt and oil at high temperatures. To recover more oil from a depleted reservoir, the foam needs to maintain a long-term

* Corresponding Author Email: huazhou@ualberta.ca.

stability in the reservoir. With the continuing innovation of nanoparticle technologies, the challenges encountered by surfactant-stabilized foam can be remedied by using nanoparticles instead of surfactant as foaming agents. Solid particles need high energy to be adsorbed to, and desorbed from the fluid interfaces; therefore, the nanoparticle-stabilized foam can be highly stable even under harsh conditions (e.g., elevated temperature and the presence of oil).

In this study, the static foam stability of nanoparticle-stabilized C_3H_8 foam is examined under different reservoir conditions. Nanocrystalline cellulose (CNC), which is naturally hydrophilic and cannot act as foam stabilizer alone, has been chosen as the nanoparticle for foam stabilization purpose. Cationic CTAB is used to modify the surface of bare anionic CNC in situ from being hydrophilic to relatively hydrophobic and surface active. We examine the effects of particle coating, surfactant concentration, salinity, temperature, and the presence of oil on C_3H_8/CNC/CTAB foam stability. It is found that the stability of CTAB-coated CNC stabilized foam increases with an increase in CTAB concentration, salinity and temperature. It is less sensitive to the presence of oil. Other foams formed by recipes of C_3H_8/CNC/Tween 20, C_3H_8/CNC/Tween 80, and C_3H_8/CNC/Triton X-100 are also tested to explore the synergistic effect on foam stability due to the use of CNC and non-ionic surfactants. It is found that, if a non-ionic surfactant is a foaming agent, the synergistic effect of CNC nanoparticle and non-ionic surfactant enhances the foam stability. In other words, stable foam can be formed by introducing both foam forming agent (surfactant) and foaming booster (nanoparticle). By comparing the stability of foam formed by CTAB-coated CNC and CNC/non-ionic surfactant mixtures, foam stability of CTAB-coated CNC stabilized foam is stronger than that of CNC/non-ionic surfactant mixtures. Finally, the dynamic mobility of foam flow through porous media is examined by alternatively injecting C_3H_8 and CTAB-coated CNC or CNC/non-ionic surfactant mixtures (SAG). We find that if a non-ionic surfactant is used as a foaming agent, the synergistic effect of CNC and non-ionic surfactant enhances the reduction in foam mobility. In addition, foam formed by alternatively injecting CTAB-coated CNC and C_3H_8 shows a higher mobility reduction effect than that formed by alternatively injecting CNC/non-ionic surfactant and C_3H_8, which is consistent with the results of the static stability tests. In the SAG tests, (either CTAB/CNC or Triton X-100/CNC stabilized C_3H_8 foam) conducted with an oil-saturated beadpack, the presence of CNC nanoparticles is able to enhance the oil recovery due to a mobility control mechanism.

Keywords: nanocrystalline cellulose, CNC, core flooding, enhanced oil recovery, nanoparticles

1. INTRODUCTION

1.1. Overview

The BP Statistical Review of World Energy shows that, at the end of 2013, total world proved oil reserves reach 1687.9 billion barrels, 60% of which are trapped in formations after primary and water flooding [1]. Such large quantity of residual oil is expected to be recovered by many enhanced oil recovery (EOR) technologies. Enhanced oil recovery, also called as tertiary recovery, is the process to reduce the residual oil saturation by gas injection, chemical injection, and thermal recovery methods. At present, gas solvent flooding technique (e.g., CO_2 or C_3H_8 based solvent) has attracted much attention in heavy oil and bitumen recovery [2, 3]. However, the high mobility ratio between the displacing gas phase and the displaced oil phase is a critical challenge encountered in solvent-flooding EOR process. In other words, such low-viscosity gas is more mobile than displaced oil or water in the reservoir; therefore, it channel through the reservoir, resulting in the early breakthrough of gas with large volume of hydrocarbon trapped in reservoir. In addition, the low-density gas fluid relative to oil will readily result in the gravity segregation, which inhibits the contact between the displacing gas and deeper hydrocarbon zones. The above problems of viscous fingering and gravity segregation are considered to be an unfavorable displacement condition; because in this case, large quantities of oil in the formation will be bypassed and the formation residue oil saturation will still be high at gas breakthrough [4]. In order to achieve a lower mobility ratio of solvent to reservoir oil, it is necessary to find a mobility control method to make the displacement sustain a uniform displacement front.

Foam has been widely used in tertiary oil recovery due to its capability of decreasing the gas-phase mobility, which can be considered as a mobility control agent. Therefore, generating foam *in situ* can be a viable approach to mitigating the sweep inhomogeneity and increase the sweep efficiency of solvent flood due to the foaming mechanism. The high apparent viscosity of foam contributes to a low mobility of the displacing fluid, i.e., gas solvent, which is critical to improving the volumetric sweep efficiency of solvent flood so as to improve the oil recovery. Conventional foam, generated *in situ* by surfactant-alternating-solvent injection (SAG), is widely used in the oil and gas industry. Surfactant is considered as an effective foam forming agent due to its amphiphilic property. In other words, the hydrophilic head of surfactant has the affinity to the water phase, while its hydrophobic tail has the affinity to gas or oil.

However, most foam generated by surfactant can be naturally unstable and thermally degradable; therefore, they cannot maintain long-term stability before fulfilling the duty, especially at elevated temperatures or with the presence of brine or oil. In addition, surfactant also has potential to be adsorbed to the rock, resulting in high surfactant retention [5]. Recently, with a growing interest in nanotechnology applications in the petroleum industry, employing the nanoparticles to replace the surfactant is an alternative way for generating foam, which can partially remedy the limitations of the traditional surfactant-stabilized foam.

1.2. Literature Review

Foam can be used to displace the trapped oil in formations, and it needs to maintain a long-term stability to fulfill its duty under the reservoir conditions.

1.2.1. Stability of Foam Formed by Surfactant

The low-density gas used for displacing oil in a reservoir hardly contacts the deeper part of the reservoir due to the gravity segregation. Its low viscosity also makes it easily channel through the formation due to a high mobility ratio between the displacing phase the displaced phase. The mobility ratio (*M*) is defined as:

$$M = \frac{\lambda_{displacing}}{\lambda_{displaced}} = \frac{\left(\frac{k_r}{\mu}\right)_{displacing}}{\left(\frac{k_r}{\mu}\right)_{displaced}} \tag{1}$$

where λ is the mobility, k_r is the relative permeability, and μ is the viscosity. Based on Equation (1), as to reduce the mobility ratio (M), i.e., overcome the gravity segregation and fingering or channeling problem, it is not practical to increase the density of displacing phase or decrease the permeability of displacing phase in the formation. The most feasible way is to increase the viscosity of the displacing phase via using foam. Extensive studies have focused on foam stability of surfactant-stabilized foam by a conventional way of solvent-alternating-surfactant method. The earliest idea of using foam to displace oil was suggested and patented by Bond and Holbrook in 1958 [6]. Soon after, Fred [7] conducted extensive laboratory experiments to further confirm that the foam can perform as a blocking agent to reduce the mobility of the displacing solvent. Following Fred's laboratory work, Bernard and Holm found that the conventional CO_2 flooding method could easily bypass large quantities of oil in the reservoir, resulting in a disappointing recovery. In 1967, they patented a notion that uses foam for mobility control in CO_2 recovery method. They noticed that the sweep efficiency was more satisfactory by just adding a small quantity of surfactant into CO_2-water flood process compared to conventional CO_2 flooding. Although the increased sweep efficiency achieved, there exists a difficulty in dissolving the Triton X-100 in CO_2 [8]. Surfactant-stabilized foam readily suffers from the surfactant degradation due to the high temperature, high salinity and the presence of oil and the surfactant absorption onto the formation rock. The use of nanoparticles to replace surfactant can overcome such limitations so as to enhance the foam stability.

1.2.2. Stability of Foam Formed by Nanoparticles

As to remedy the lack of the capability of providing stable foam over a long period by surfactant under harsh reservoir conditions, the surfactant can be replaced by nanoparticles to generate foam with outstanding foam stability. Colloidal particles on the nanoscale, behaving like surfactant, can act as a foam stabilizer by spontaneously residing at fluid interfaces. Nanoparticles can be irreversibly attached to the immiscible gas-liquid or liquid-liquid interfaces due to their high adsorption energy. Such high adsorption energy of nanoparticles might enable the foam to tolerate challenging conditions including high temperature, high salinity, and the presence of oil.

In order to achieve long lasting foams, nanoparticles need to exhibit a certain contact angle at the interfaces, indicating a strong free energy of adhesion to the fluid interfaces [9, 10]. As for spherical particles, the minimum free energy of particle detachment (*E*) is given as:

$$E = \pi r^2 \gamma (1 \pm cos\theta)^2 \quad (2)$$

where r is spherical particle radius; γ is interfacial tension; and θ is contact angle. Equation (2) shows that the particle detachment energy increases as contact angle increases until it reaches 90°, and then decreases with a further increase in contact angle. When the contact angle is around 90°, nanoparticles can be irreversibly trapped into the liquid interface with an energy thousand times of $k_B T$. As for non-spherical particles, the particle orientation and aspect ratio need to be considered in the detachment energy. Binks and Horozov [10] also reported that particles with a disklike-shape exhibit a higher detachment energy than those with a rodlike-shape, and rodlike-shape particles have a higher detachment energy than the spherical particles.

There are extensive new researches associated with the nanoparticle-stabilized foams. The pioneering idea of nanoparticle-stabilized foams or emulsions, also called as pickering emulsion, was suggested by Ramsden [11] and Pickering [12] in the early twentieth century. Studies by Alargova et al. [13] demonstrated that the hydrophobic polymer microrods can generate super-stabilized aqueous foams, which can keep a constant foam volume for many days, in contrast to a few hours' lifetime of foam generated by sodium dodecyl sulfate (SDS) surfactant. Soon after, Binks and Horozov [14] hydrophobized the fumed silica to test the foam stability with varied particle hydrophobicity. Their findings showed that particles with an intermediate hydrophobicity are surface active and tend to have a higher chance of being attached to the fluid interfaces. However, most raw nanoparticles are inherently hydrophilic; thus it is necessary to modify the wettability of these colloidal particles from being hydrophilic to being more hydrophobic. Espinoda et al. [15] obtained very stable in-situ supercritical CO_2-in-water foams by co-injecting the CO_2 and the dispersion of hydrophilic fumed silica coated with polyethylene glycol (PEG) through a glass-beads pack with 180 um glass beads. The resistance to flow calculated from the measured differential pressure of the capillary tube showed that the viscosity of silica-stabilized foam is two to eighteen times than fluids without nanoparticle. Later on, Yu et al. [16] studied the effect of total injection rate and phase ratio of CO_2/silica dispersion on the foam mobility through glass beads. The results showed that the higher the total flowrate and phase ratio of CO_2/silica dispersion, the higher the foam mobility. Worthen et al. [17] employed the partially hydrophobic silica modified by methyl silyl ligands to achieve a long-term CO_2 in water foam, which revealed greater foam stability than that formed by PEG-coated silica.

However, surface treatment of non-surface-active nanoparticles through chemical modification may be expensive and complicated. Later, some researchers found that altering the particle surface via physiochemical adsorption is also an effective method. Zhang et al. [18] reported that the combination of inherently hydrophilic laponite clay with non-ionic surfactant $C_{12}E_4$ created a synergistic effect on stabilization of foam. Recently, static tests by Singh et al. [19] revealed that foam stabilized by mixture of silica and anionic surfactant is more stable at elevated temperature and with the presence of oil than the surfactant-stabilized foam.

Up to date, most previous works were focused on determining experimentally the foam stability generated by N_2 or CO_2; however, C_3H_8 is also a commonly used rich gas in the solvent flooding, none of the researches have focused on C_3H_8 stabilized foam. The gas type may have a significant effect on the foam stability. In addition, most of the previous studies only focused on exploring the foam stabilization mechanism of spherical nanoparticles. No research has been conducted on foam stabilized by one unique type of non-spherical nanoparticles, i.e., cellulose nanocrystals (CNC). CNC is derived from low-cost, environmental-friendly and abundant fibre materials. Besides, only a few of studies examined the foam stability of foam generated either by surfactant or nanoparticles under harsh reservoir conditions, such as high salinity, elevated temperature, and presence of oil [15, 16, 17, 18]. Therefore, it is highly necessary to test the foam stability of foam generated by C_3H_8 in both conventional (surfactant-stabilized foam) and nonconventional (nanoparticle-stabilized foam) way under reservoir conditions.

1.3. Objectives

The objective of this study is to systematically test the static stability and dynamic mobility of C_3H_8 foam stabilized by nanoparticle experimentally by using a pressure/volume/temperature (PVT) setup and a glass beadpack, respectively. Static foam stability test is conducted by recording the foam height as a function of time. Dynamic mobility test is conducted by recording the pressure drop across the porous glass beadpack. The detailed objectives are listed as follows,

1) To test the static foam stability of C_3H_8 foam stabilized by silica/Triton X-100, CTAB modified CNC, or CNC/nonionic surfactant mixture. The influencing factors including concentration, salinity and temperature are considered. The effect of presence of oil on foam stability is only considered for the CNC/CTAB system.
2) To test the mobility of NAG-generated foam (nanoparticle alternating gas) by allowing the foam to flow through a porous glass beadpack.

2. METHODS

2.1. Materials

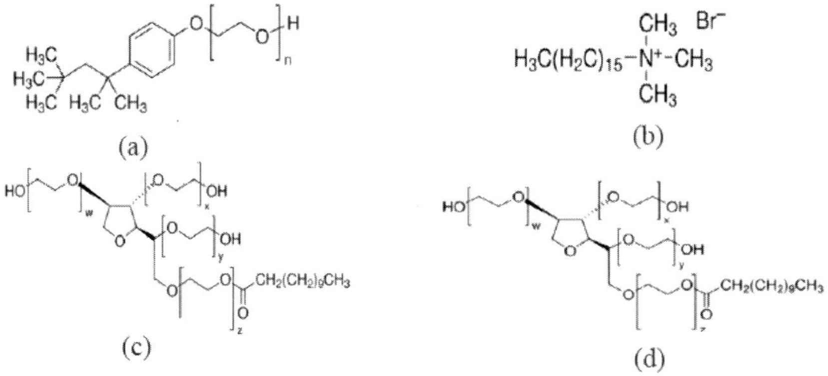

Figure 1. Molecular structures of (a) Triton®X-100 surfactant with $n = 10$; (b) CTAB surfactant; (c) Tween 20 surfactant; (d) Tween 80 surfactant (Supplied by Sigma-Aldrich).

CNC is fibrous anionic-type nanocrystalline cellulose derived from wood pulp, which potentially offers environmental advantages than other nanoparticles. CNC is hydrophilic in nature. Triton X-100 (See Figure 1(a) for its molecular structure with n=10) is the non-ionic polyoxyethylene octylphenyl ether surfactant with a purity of 100%, purchased from Sigma-Aldrich. CTAB (See Figure 1(b)), also called as hexadecyltrimethyl-ammonium bromide, is a cationic surfactant, supplied by Sigma-Aldrich in a powder phase with >=99% purity. Nonionic surfactant Tween 20 (See Figure 1(c)) and Tween 80 (See Figure 1(d)) with purity of 100% are purchased from Sigma-Aldrich. Solid sodium chloride (NaCl) with >99% purity is provided by Acros Organic. Liquid C_3H_8 (Praxair) with 99.5% purity is depressurized into the gas phase in PVT cell prior to foam generation. To investigate the effect of heavy oil presence on foam stability, n-$C_{16}H_{34}$ with 99% purity is chosen as the pseudo-oil to simulate heavy oil for studying. The nanoparticles, surfactants and NaCl are all weighed by an electronic balance and their concentrations are expressed in w/v (weight over volume fraction) in the aqueous phase.

2.2. Experimental Setup

2.2.1. Foam Stability

The mercury-free DBR PVT system (Schlumberger, Edmonton, Alberta, Canada), as shown in Figure 2, is the experimental apparatus used to test foam stability. Embedded in a PVT system is a transparent PVT cell, which can withstand pressures up to 15000 psi. The cell temperature, controlled by the microprocessor temperature controller fitted in the air bath, can tolerate a temperature range of -10°C to 200°C. The cell is separated into two isolated chambers by a mobile piston, above and below which are a hydraulic oil chamber and a sample chamber, respectively. The sample chamber volume is changeable by moving the piston up and down through a positive displacement pump (DBR pump). The bottom of the sample chamber is equipped with a magnetic stir that can stir the sample at different rates. Two charging vessels are used to transfer the solution samples and *n*-$C_{16}H_{34}$ into the PVT cell. A Vacuum Pump (1400, Welch Vacuum, USA) is used to vacuum the PVT cell. A cathetometer with a resolution of 0.002 cm is used to measure the fluid height in the visual cell.

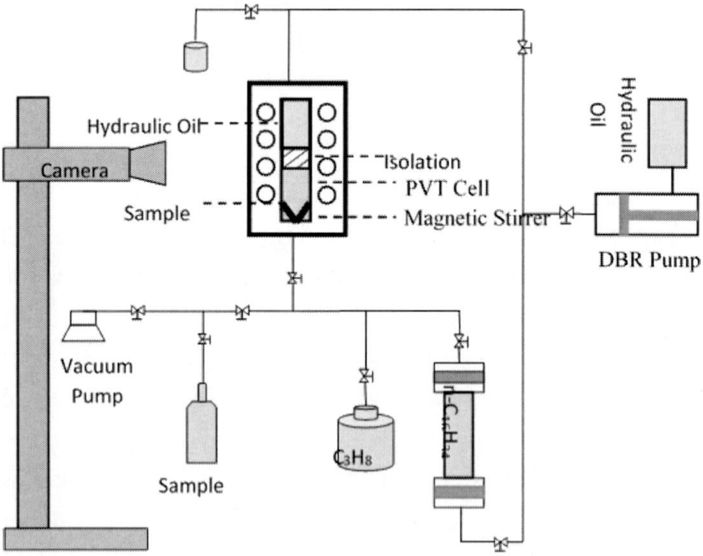

Figure 2. Schematic of the PVT experimental setup used for foam stability tests.

2.2.2. Foam Mobility

Figure 3 shows a diagram of the experimental apparatus used to test the mobility of foam generated using the NAG method. A 38.400-cm long stainless steel tube with 0.386 cm inner diameter has been packed tightly with 177 μm to 250 μm spherical glass beads. Glass beads are sealed in place by metal screens attached to the tubing ends. Surfactant solution or nanoparticle dispersion is first placed in a transfer cylinder and then injected by a syringe pump at a constant rate into the beadpack. C_3H_8 gas is injected from a gas tank, the flowrate of which is controlled by a mass flow controller (RK 32907-57, Cole-Parmer, Canada). The dynamic pressure difference is measured by a differential pressure gauge (PM, Heise, US) that is connected to the two ends of the glass beadpack.

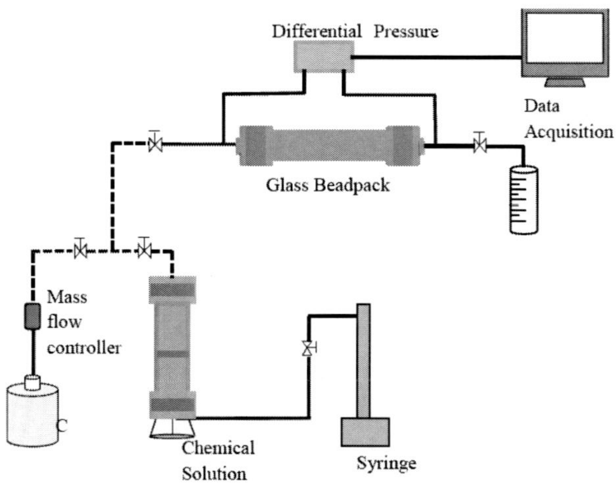

Figure 3. Schematic of the experimental setup for studying foam flow through porous media.

2.3. Experimental Procedure

2.3.1. Preparation of Nanoparticles Dispersion

Each sample is freshly prepared prior to each test. To prepare the aqueous dispersion, 0.5 w/v bare silica or 0.3 w/v bare CNC are prepared by dispersing the particles in de-ionized water for 40 min by using an

Ultrasonifier (FS30, Ultrasonic Cleaner, US). To prepare the silica/Triton X-100 mixture, 0.0194 w/v Triton X-100 is added into the silica dispersion. Since anionic CNC is hydrophilic due to the presence of hydroxyl groups, a cationic surface modifier, CTAB, with 0.002 w/v, 0.010 w/v, or 0.015 w/v, is added into bare CNC dispersion to change the wettability of CNC from being hydrophilic to more hydrophobic, as shown in Figure 4. To prepare CNC/nonionic surfactant mixtures, Tween 20, Tween 80, and Triton X-100 are added into the bare CNC dispersion, respectively, at their CMCs. Sodium chloride solution is prepared at desired concentrations and mixed into nanoparticle solutions. All the above particle/surfactant dispersions are stirred for 30 min to ensure the mixture homogeneity.

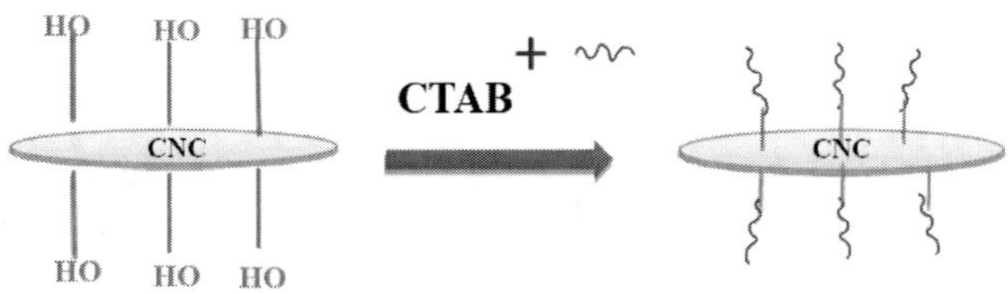

Figure 4. Schematic showing how to perform surface modification of bare CNC by CTAB.

2.3.2. Foam Stability Determination

The PVT cell and tubing are carefully cleaned with toluene and de-ionized water, followed by an evacuation by a vacuum pump before each test. As for temperature control, the air bath temperature is set at a desired value, followed by at least 12 hours to ensure a constant temperature. Freshly prepared 15 cc of solution sample is initially transferred into PVT cell via the charging vessel. Then C_3H_8 is injected from the bottom of the sample cell at a constant flowrate of 500 cc/hr. The volume ratio of C_3H_8 gas to sample solution is kept as 2:1 throughout all tests. Right after the gas sparging terminates, static foam can be generated by stirring the solution for 2 min with the stirrer. Then quantitative evaluation of the foamability is achieved by measuring the initial foam height just after the mixing terminates. The foam stability can be evaluated by recording the foam height as time elapses. In the tests with the presence of oil, 5 cc of oil is introduced into PVT cell at a rate of 500 cc/hr exactly after the mixing terminates. The bubble size and foam uniformity can be observed with naked eyes through PVT cell during the tests.

Transmission electron microscope (TEM) (Morgagni 268, Philips) and optical microscope are used to observe the particle shape, distribution, and bubble structure. In this case, the foam is formed by handshaking the solution with air, since it is difficult to collect C_3H_8 foam out of the PVT cell. TEM imaging is carried out as follows. First, a small volume of sample foam is placed on 1 cm diameter stub and dried in an oven. Then the sample is transferred into TEM chamber, followed by a vacuum process. Then TEM image can be acquired.

2.3.3. Foam Mobility Determination

The beadpack is refilled with glass beads before each test. It is then flooded by 15 PV of the tested solution; the beadpack's permeability is observed to be almost the same in all tests. Initially, C_3H_8 and water are alternately injected into the system (WAG) with an injection period of 2 min for each stage, which is used as a baseline test. The injection rates for C_3H_8 and water are 6 cm^3/min and 3 cm^3/min (WAG ratio of 1:2), respectively. Then the experiments are repeated by alternately injecting C_3H_8 and CTAB-coated CNC or alternately injecting C_3H_8 and CNC/non-ionic surfactant mixture (SAG) under the same conditions as used in baseline WAG test. The pressure difference across the beadpack, recorded during the experiments, is used to estimate the mobility of foam flowing through porous media. All the aforementioned WAG and SAG tests are conducted at 20°C.

In the tests with the presence of oil, a newly packed beadpack is firstly saturated with n-$C_{16}H_{34}$. Then surfactant solution or nanoparticle dispersion is alternately injected with C_3H_8 gas into the beadpack at the same injection conditions as used in the baseline WAG test. The duration of either chemical solution injection or C_3H_8 gas injection is kept as 2 min. The changes of differential pressures over time are monitored during the SAG process. Several graduated cylinders are used to collect the oil and solution at the outlet of the beadpack, in order to record how much oil has been produced. All the above experiments are conducted at 20 °C.

Table 1. Properties of the glass beadpack used in the experiments

Length (cm)	38.400
Diameter (cm)	0.386
Pore volume (cm^3)	1.823
Porosity	40.58%
Permeability (Darcy)	34.16

During the tests, pressure differences are recorded as a function of time. The permeability of the beadpack is measured to be 34.16 Darcy by recording the steady-state pressure differences when water is flowing through the beadpack at different rates. Other properties of the glass beadpack are detailed in Table 1. The fluid mobility can be calculated by applying Darcy's law:

$$\lambda = \frac{qL}{A\Delta p} \tag{3}$$

where λ is the fluid mobility in Darcy/cp; q is the volumetric flow rate of fluid in cm^3/s; L is the beadpack length in cm; A is the cross-sectional area of the beadpack in cm^2; Δp is the differential pressure along the beadpack in atm. All these foam mobility tests are conducted at the room temperature of 20°C.

3. Results and Discussion

3.1. Foam Stabilized by CNC Hydrophobized with CTAB

3.1.1. Effect of Surface Coating on Foam Stability

Several bare CNC solutions with concentrations of 0.05 w/v, 0.3 w/v, and 0.5 w/v are mixed with C_3H_8 to generate foam at 20°C and 700 kPa; however, no foam is generated by CNC alone without addition of surfactant as a surface modifier. Based on the previous study [10], particle can act as a foam stabilizer only when wettability of nanoparticle surface is partially hydrophobic and partially hydrophilic, which allows the particles to disperse in water due to the hydrophilic property and to be adsorbed at liquid interfaces due to the hydrophobic property. The formation of foam requires adsorption of solids at air/water interface; therefore, it is necessary to select a suitable surface modifier that can render the CNC surface be modified from being hydrophilic to partially hydrophobic via either electrostatic or ligand exchange reactions. Since CNC is negatively charged, the cationic hydrophobic surfactant CTAB can be coated onto its surface via electrostatic interaction to modify the surface wettability of CNC particles, as shown in Figure 4. Here, CNC with a constant 0.3 w/v concentration is coated by adding surface modifier CTAB with concentrations of 0.02 w/v, 0.010 w/v, and 0.015 w/v, respectively, at 20°C and 700 kPa. The foamability and foam stability of CNC/CTAB-stabilized foams are compared with CTAB-stabilized foam at the same conditions. Table 2 and Figure 5 shows the foam stability test results for foam stabilized by 0.3 w/v CNC coated with different CTAB concentrations. It shows an encouraging result that the coated CNC increases the foam stability. No foam can be formed solely by 0.002 w/v CTAB surfactant, while a short and stable column of foam can be formed when 0.3 w/v CNC is coated by 0.002 w/v CTAB in aqueous phase. This indicates that in-situ surface-coated nanoparticles are responsible for such foam generation. As shown in Figure 6, at increased CTAB concentrations (0.010 w/v and 0.015 w/v), foam prepared from CTAB solutions alone collapses completely within 12 hours; however, CNC/CTAB stabilized high-volume foam can retain 4/5 of the initial height after 12 hours. Images of foam columns (Figure 6) are captured, 3 hours after the mixing terminates, at different concentrations of CTAB with and without the presence of CNC. It is found that, 3 hours after the mixing terminates, the coated CNC/CTAB foam exhibits a much thicker column than CTAB-stabilized foam. Figure 5 also compares the decay profiles of foams generated by CNC/CTAB and CTAB alone, respectively. CNC/CTAB-stabilized foam decays more gently in contract to the CTAB foam that shows a remarkable drop in foam height as time elapses. This is attributed to that surfactant-stabilized foam is easy to attach to and detach from the liquid interface; on the contrary, CNC-stabilized foam can be irreversibly attracted onto liquid interfaces due to its high adhesion energy [10]. Therefore, CNC particle-stabilized foam has a high adhesion energy that prevents itself from destabilization.

Table 2. Foamability and foam stability of foams formed by CTAB and 0.300 w/v CNC/CTAB at 20°C. No foam could be generated solely by 0.300 w/v CNC dispersion

CTAB concentration, w/v	Surfactant			Surfactant + CNC			Mixed dispersion
	Initial foam height, cm	H_{12}, cm	$T_{1/2}$, min	Initial foam height, cm	H_{12}, cm	$T_{1/2}$, min	
0.002	0	0	0	0.539	0	245	Stable foam
0.010	1.237	0	240	1.035	0.830	>1000	Stable foam
0.015	2.123	0	280	2.885	2.310	>1000	Very stable foam

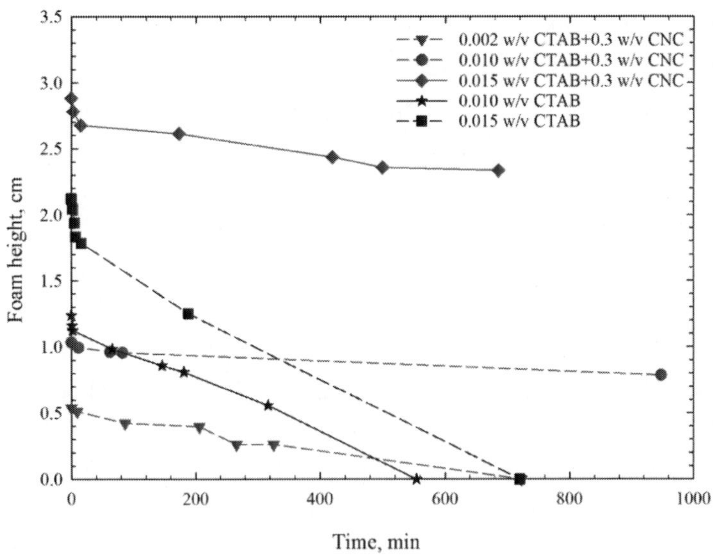

Figure 5. Effect of CTAB concentration on the stability of the CTAB/CNC-stabilized foam and CTAB-stabilized foam (20°C, 700 kPa).

It is also found that (Table 2 and Figure 5), an increase in CTAB concentration from 0.002 w/v to 0.015 w/v leads to an increased foamability and foam half-life of CNC/CTAB foam. This is probably because: at lower surfactant concentrations, there are insufficient particle surfaces to be coated with hydrophobic surfactant, resulting in a relative low foamability compared to the high surfactant concentration cases. As surfactant concentration increases, more particle surfaces are altered to be partially hydrophobic; therefore strong attachment occurs onto interfaces, contributing to an increased foaming capability.

Macroscopic bubble characteristics are observed by naked eyes. In the presence of CNC particles, an uniform foam is observed in the PVT cell. The opaque white bubbles have a needle tip-like size with a diameter smaller than 1 mm. They are evenly packed in the beginning and maintain almost the same size (1 mm to 1.5 mm) and distribution as time elapses. However, in the absence of CNC particles, narrow packing of fine bubbles with a diameter smaller than 1 mm is observed initially; but the bubbles gradually coarsen to 2-3 mm as time elapses. The size of bubbles stabilized by CNC does not vary much as time goes by, which is consistent with the previous reports that particles can slow down the bubble coarsening process [14]. Bubble coarsening is one type of foam destabilization mechanism. When the gas pressure inside the small bubbles exceeds the pressure inside the large bubbles, gas will diffuse into the large bubbles from small bubbles through the liquid film, resulting in the bubble coarsening and consequently bubble rupturing [20]. Hence, our study confirms that nanoparticle-stabilized foam is much more stable compared to the conventional surfactant-stabilized foam.

Figure 7 shows the microscopic examination applied to further quantify the size and distribution of both 0.010 w/v CTAB-stabilized foam and 0.3 w/v CNC/0.010 w/v CTAB-stabilized foam at ambient temperature and pressure. It is noticed that the average size of foams generated by CTAB alone is two to three times larger than that generated by coated CNC nanoparticles. Most CTAB-stabilized foams are in donut-like shape with a diameter of 220 μm, while CNC stabilized foams are mostly 60 μm and 180 μm in diameter.

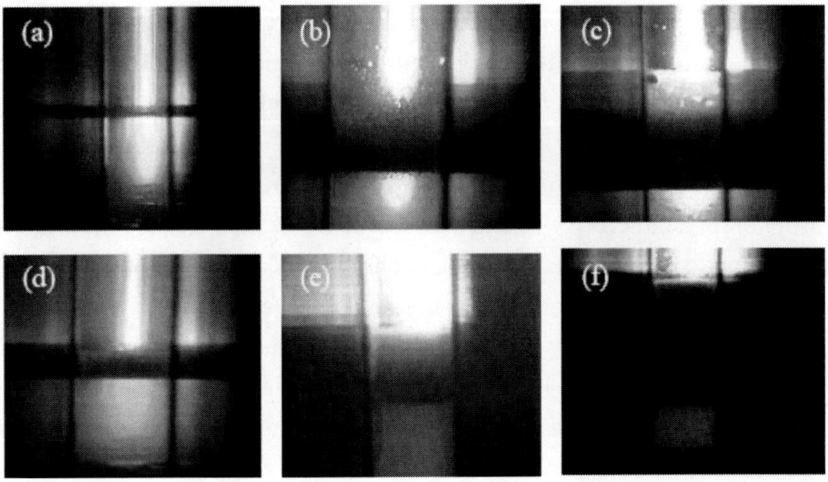

Figure 6. Digital images of foam stabilized by (a) 0.002 w/v CTAB (no foam formed); (b) 0.010 w/v CTAB; (c) 0.015 w/v CTAB; (d) 0.300 w/v CNC coated by 0.002 w/v CTAB; (e) 0.300 w/v CNC coated by 0.010 w/v CTAB; (f) 0.300 w/v CNC coated by 0.015 w/v CTAB (3 hours after the mixing was terminated, at 20°C and 700 kPa).

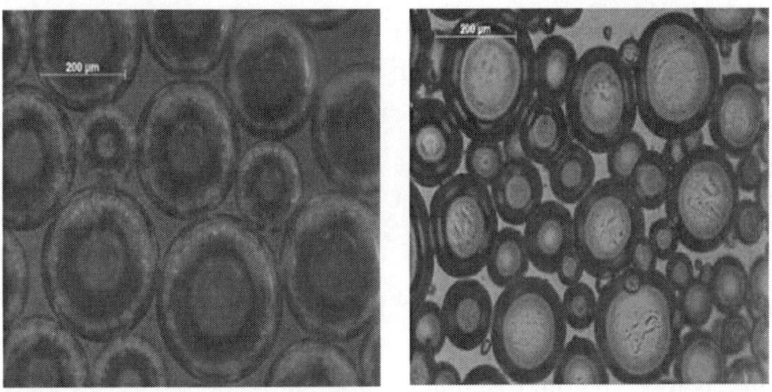

Figure 7. Optical microscopic images of aqueous foams stabilized by 0.010 w/v CTAB alone (left) and 0.300 w/v CNC coated by 0.010 w/v CTAB (right). Scale bars are given separately.

Figure 8 shows that the fibrous bare CNC particles are dispersed in water, but coated CNC particles align along the films of foams by TEM imaging. For the bare CNC system, no foam is formed, so the CNC nanoparticles are randomly distributed. As for the coated CNC system, large quantities of foams are formed, and CNC nanoparticles are found to aggregate at the bubble films.

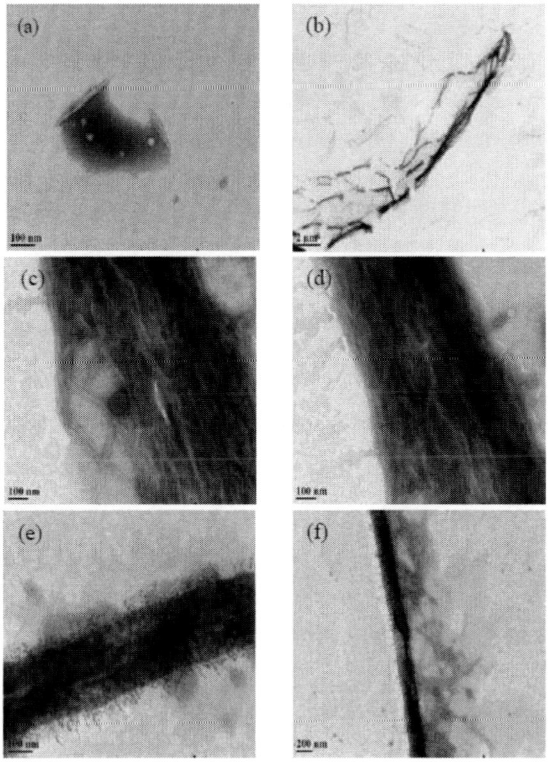

Figure 8. TEM images of (a) fibrous bare CNC dispersed in water; (b) (c) (d) dry and curved liquid film of bubbles with coated-CNC particles aligned along; (e) (f) another dry and curved liquid film of bubbles with coated-CNC particles aligned along. The CNC dispersion (a) was formed by using 0.300 w/v CNC particles alone; the others were formed by using 0.300 w/v CNC coated by 0.010 w/v CTAB. Scale bars are given separately.

3.1.2. Effect of Salt Concentration on Foam Stability

To investigate the salinity effect on foam stability, 10000 ppm of NaCl is added to the mixture of 0.3 w/v CNC and 0.010 w/v CTAB dispersion at 20°C and 700 kPa. Table 3 shows the initial conditions and test results on the foam formed by CTAB coated CNC under high brine concentrations. It is interesting to note that, compared with the case without salt, the presence of 10000 ppm salt increases the initial foam height from 1.035 to 2.803 cm but with a 0.560 cm column of flocculates below the foam column. The height of flocculates keeps a constant 0.560 cm during the foam decay. Such result is consistent with previous tests, which shows that fumed silica with 67% and 80% SiOH can stabilize foam when salt is present in the dispersion [21]. With the presence of salt, a thick foam column is obtained, which is probably because the addition of monovalent ion neutralizes charges and increases hydrophobicity of particles, thus promoting the adsorption of CNC onto the liquid interfaces for generating more foam. The addition of salt to the solution not only increases the foam thickness but also results in finer foam texture with a uniform packing.

Table 3. Foamability and foam stability of foams formed by 0.300 w/v CNC coated with 0.010 w/v CTAB with addition of salt at 20°C

Brine concentration, ppm	Surfactant + CNC			Mixed dispersion
	Initial foam height, cm	H12, cm	T1/2, min	
0	1.035	0.830	>1000	Stable foam
10000	2.803	2.306	>1000	Very stable foam (flocculate below the foam)

Figure 9 shows the foam decay profiles as a function of time for the tests with and without salt presence. Although the foam without the presence of salt exhibits a gentle and smooth slope, the foam without the presence of salt shows an even milder slope. The higher initial height and gentle decay trend of foam with the presence of salt indicate a higher foamability than the case without salt presence. As mentioned by Deschenes [22], there exists a hydrophobic force to rupture the liquid film. Therefore, a decrease in the hydrophobic force can result in an increase in foam stability. Craig [23] reported that some common salts can inhibit bubble coalescence. He measured the bubble coalescence rate of foams formed by different electrolytes and found that the hydrophobic force decreased with an increase in the electrolyte concentration. It can be concluded that the presence of salt concentration might decrease the hydrophobic force on bubble coalescence, resulting in an increase in foam stability. Digital photographs are taken to record the change in foam height and foam texture over time. Figure 10 illustrates the foam texture generated with the addition of electrolyte concentration, showing a thicker foam column with larger bubble density and smaller bubbles. During the foam decay process, macroscopic observation reveals that with the addition of 10000 ppm salt into the dispersion, the bubble size is homogeneous and smaller than 1 mm initially, and increases to around 1 mm to 2 mm after 12 hours.

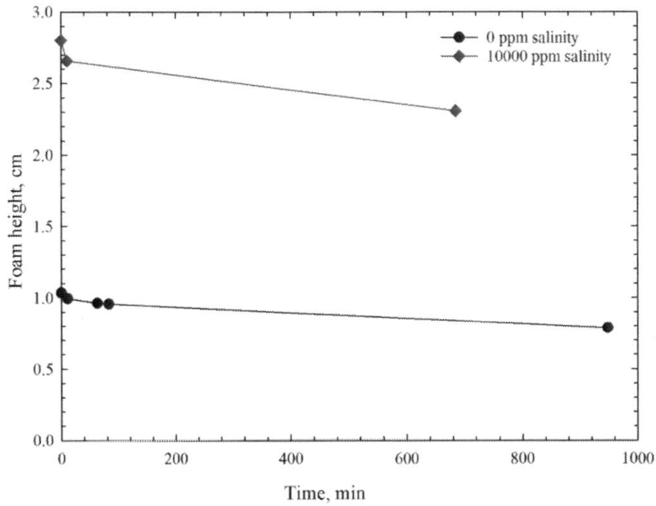

Figure 9. Effect of salinity on the stability of foam formed by CNC/CTAB (0.010 w/v of CTAB, 0.300 w/v of CNC, at 20°C and 700 kPa).

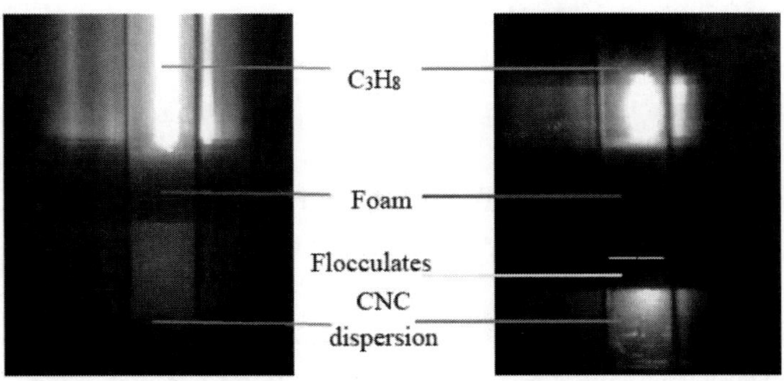

Figure 10. Digital images of CNC/CTAB stabilized foams (0.300 w/v CNC and 0.010 w/v CTAB) with (right) and without (left) adding 10000 ppm salt, respectively, at 20°C and 700 kPa, 5 hours after the mixing was terminated.

3.1.3. Effect of Temperature on Foam Stability

Foam is a thermodynamically unstable fluid, so it is worth of testing the foam stability formed by CTAB coated CNC at elevated temperatures. To investigate the temperature effect on foam stability, a series of experiments are conducted at temperatures of 20°C, 50°C, and 80°C, respectively, but at 700 kPa by using 0.3 w/v CNC coated by 0.010 w/v CTAB or 0.010 w/v CTAB alone. Table 4 shows the foamability and foam stability for foams formed by CTAB and CTAB coated CNC at elevated temperatures. In the absence of CNC, foam solely formed by CTAB surfactant disappears completely within 12 hours at 20°C and 50°C (Table 4). No foam is observed as temperature increases to 80°C. It can be concluded that CTAB stabilized foam is very sensitive to temperature and the stability decreases with an increasing temperature. It is widely accepted that, since foam is thermodynamically unstable, a higher temperature tends to reduce its interfacial area and interfacial free energy via bubble coalescence [24]. Conversely, it is interesting to note that, in the presence of CNC, the initial height and the foam height after 12 hours at 20°C is almost the same as those at 50°C. When temperature increases to 80°C, the foam becomes more stable with an initial height of 2.197 cm and a remaining height of 1.328 cm after 12 hours. It can be speculated that the adhesive strength of CNC particle and CTAB surfactant increases as temperature increases. To summarize, CTAB coated CNC foam keeps almost the same stability at 20°C and 50°C, while the foam stability increases as temperature increases from 50°C to 80°C. On the contrary, foam stability of CTAB stabilized foam decreases with an increase in temperature.

Half-life, the time it takes for the foam to decay to half of the initial height, is an alternative way to quantify the foam stability. The half-life of CTAB foam is 240 min at 20°C and 15 min at 50°C, while the half-life of CTAB coated CNC foam at 20°C, 50°C and 80°C are all greater than 1000 min (Table 4). It is thus proven that surface modified nanoparticles can generate stable foams that are so strong that a high tolerance to temperature can be achieved. Figure 11 shows the evolution of foam height as a function of time at 3 different temperatures for both cases.

Table 4. Foamability and foam stability of foams formed by 0.010 w/v CTAB alone and 0.300 w/v CNC coated with 0.010 w/v CTAB at elevated temperatures

Temperature, °C	Surfactant			Surfactant + CNC			Mixed dispersion
	Initial foam height, cm	H12, cm	T1/2, min	Initial foam height, cm	H12, cm	T1/2, min	
20	1.237	0	240	1.035	0.830	>1000	Stable foam
50	0.691	0	15	1.069	0.849	>1000	Stable foam
80	0	0	0	2.197	1.328	>1000	Stable foam

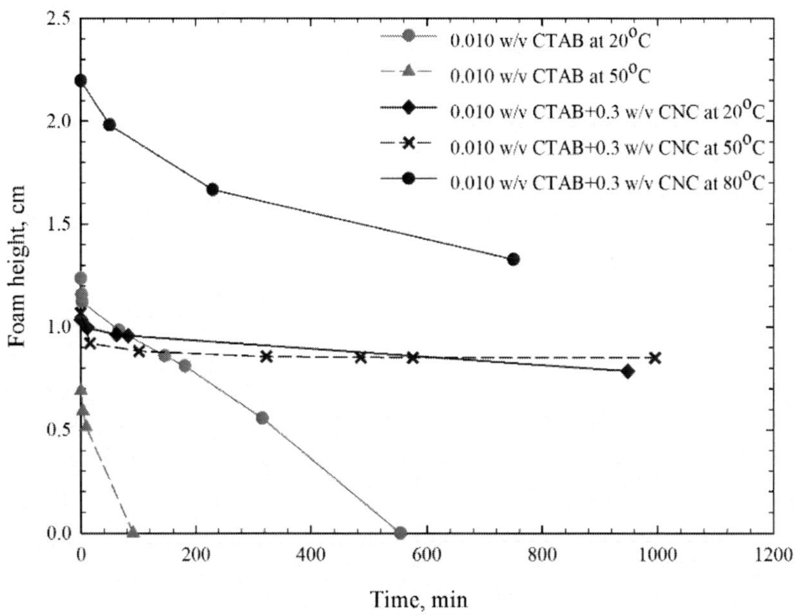

Figure 11. Effect of temperature on the stability of foam generated by CTAB (0.010 w/v) and CNC/CTAB (0.300 w/v CNC coated by 0.010 w/v CTAB), respectively.

It is clearly observed that the decay trend of foam stabilized by surfactant alone is much steeper; as compared to the nearly horizontal foam height stabilized by CTAB coated CNC. Specifically, at 50°C, the foam formed by CTAB coated CNC remains a 0.893 cm height after 85 min, while the foam formed by CTAB alone completely disappears in 85 min.

3.1.4. Effect of Presence of Oil on Foam Stability

Since foam needs to be stable to displace oil under the reservoir conditions, the presence of oil on foam stability is worth of study. Extensive studies showed that the presence of oil could diminish the stability of surfactant/CO_2 foam [25]. Therefore, it is highly necessary to test the stability of foam stabilized by C_3H_8/CTAB coated CNC with the presence of oil. As the oil droplets replace the adsorbed surfactant on the interface, it reduces the surface elasticity of the foam [26]. In our previous tests, we tested the foam stability generated by C_3H_8/Triton X-100 with the presence of oil and found that the oil was detrimental to the foam stability since C_3H_8 was much more soluble in oil than CO_2 [27]. Tests are conducted by using 0.03 w/v CNC coated with 0.015 w/v CTAB and 0.015 w/v CTAB alone at 20°C and 700 kPa, with and without the addition of pure pseudo-oil *n*-$C_{16}H_{34}$. Figure 12 shows the foam decay profile of foams formed with and without the addition of oil. As illustrated in Figure 12, for both systems, foam stability is diminished by the addition of oil. Oil can be considered as a film breaker as it enters the gas/liquid interface and spreads over the bubble film, resulting in film rupturing [26]. By comparing the foams stabilized by CNC/CTAB and by CTAB alone, with the presence of oil, CNC/CTAB foam exhibits about 1000 min of half-life, which is 25 times of that for CTAB foam. This is probably because that the presence of CNC particles at the gas/liquid interface slows the process of oil droplets entering and spreading over the bubble film.

Figure 13 shows the image captured 8 hours after the introduction of oil to the CNC/CTAB foam, showing that the foam height decreases with the addition of oil. It can be observed that as soon as the oil comes in contact with the bottom part of foam, the oil spreads over the lamellas of foam, resulting in an expedited rupturing of foam with time. These results indicate that the CNC-stabilized foam is sensitive to oil, but such foam is still very stable with a half-life of 1000 min.

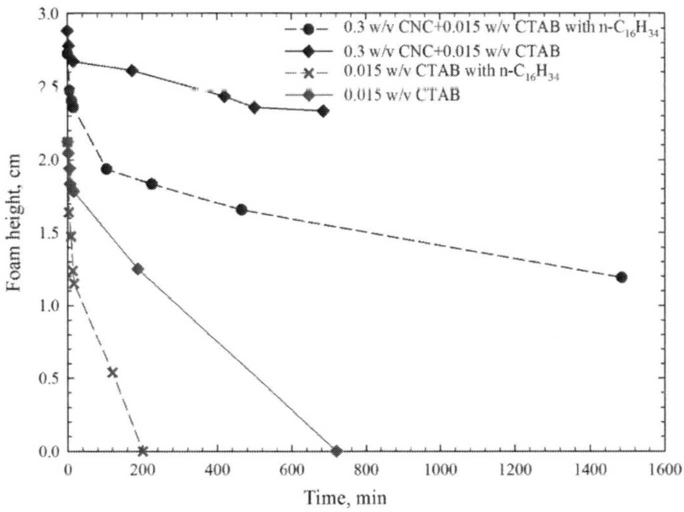

Figure 12. Effect of oil presence on the foam stability of foam generated by CTAB alone (0.015 w/v) and CNC/CTAB (0.300 w/v CNC coated by 0.015 w/v CTAB), respectively, at 20°C and 700 kPa.

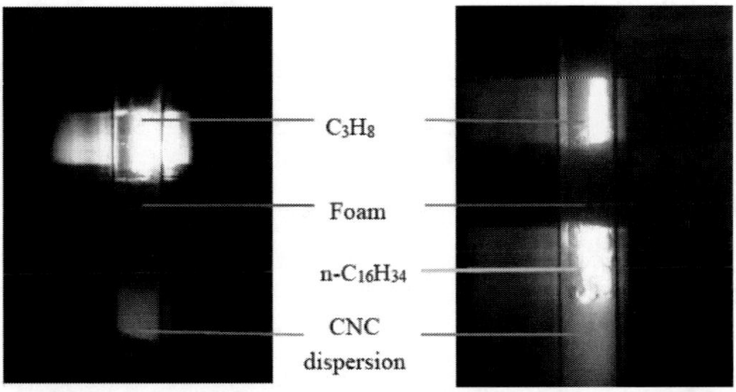

Figure 13. Digital images of CNC/CTAB stabilized foams (0.300 w/v CNC coated with 0.015 w/v CTAB) with (right) and without (left) presence of oil, respectively, at 20°C and 700 kPa, 8 hours after the mixing was terminated.

3.2. Stability of Foam Generated by Non-Ionic Surfactant and CNC

Nonionic surfactants Tween 20, Tween 80 and Triton X-100 at their CMCs are used, respectively, to mix with 0.3 w/v CNC dispersion to form a foam at 20°C. Another three foam stability tests are conducted by using these three surfactants alone, without adding CNC particles. Foams stabilized by these nonionic surfactants and surfactant/CNC are compared at the same conditions. Table 5 and Figure 14 show the test results on foam stability of foam formed by different combinations of surfactant with CNC at different temperatures. No foam is formed by Tween 80 and Tween 80/CNC at 20°C. The foam formed by Tween 20/CNC mixture shows a half-life two times longer than that formed by Tween 20 alone at 20°C, which indicates that the addition of CNC particles in the nonionic surfactant solution increases the foam stability. No foam is generated by Tween 20 alone at 50°C, while there is a short column of 0.596 cm of foam formed by Tween 20/CNC mixture at the same condition. The Triton X-100/CNC foam also exhibits a longer half-life than the Triton X-100 foam tested previously at both 20°C and 50°C [27]. All these results add up to a conclusion that, if a surfactant can generate foam itself, then the surfactant/CNC mixture can form an even more stable foam under the same conditions.

Table 5. Foamability and foam stability of foams formed by different combinations of nonionic surfactant with 0.3 w/v CNC dispersion at different temperatures

Type of surfactant	Surfactant concentration, w/v	With CNC	Temperature, °C	Initial foam height, cm	T1/2, cm	Mixed dispersion
Tween 20	0.0071 (CMC)	without	20	0.677	360	Foam
Tween 20	0.0071 (CMC)	with	20	0.957	1179	Foam
Tween 20	0.0071 (CMC)	without	50	0	0	No foam
Tween 20	0.0071 (CMC)	with	50	0.596	33	Foam
Tween 80	0.0019 (CMC)	without	20, 50	0	0	No foam
Tween 80	0.0019 (CMC)	with	20, 50	0	0	No foam
Triton X-100	0.0194 (CMC)	without	20	3.688	585	Foam
Triton X-100	0.0194 (CMC)	with	20	2.627	1250	Foam
Triton X-100	0.0194 (CMC)	without	50	2.823	12	Foam
Triton X-100	0.0194 (CMC)	with	50	2.779	121	Foam

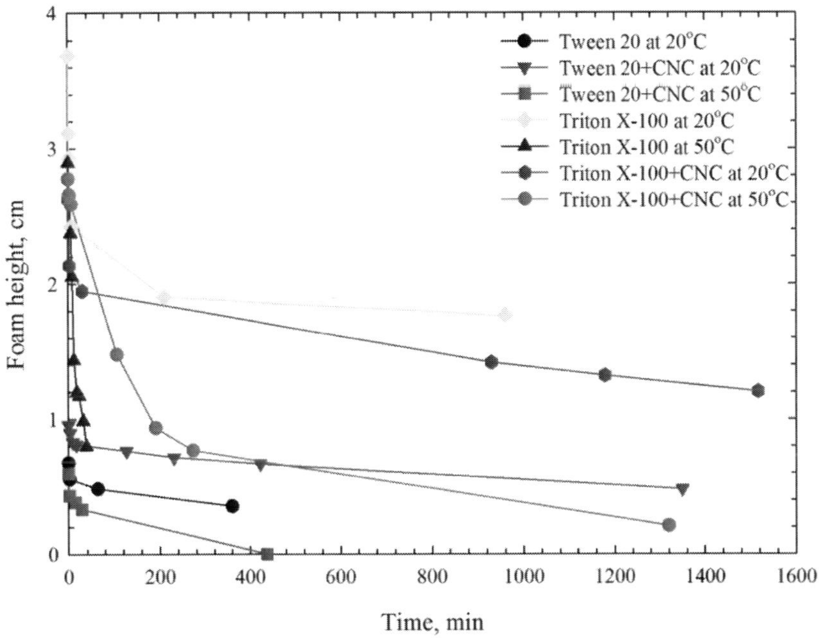

Figure 14. Comparison of stability of foams stabilized by a mixture of CNC with different nonionic surfactants (0.300 w/v CNC with surfactant at their CMCs, 20°C and 700 kPa).

According to Table 5, the initial heights of foam generated by Triton X-100 and Triton X-100/CNC are higher than those formed by Tween 20 and Tween 20/CNC mixture. This indicates that the foamability of the surfactant/CNC mixture is proportionally related to the foamability of surfactant alone. In other words, the nonionic surfactant can be considered as a foaming agent, and CNC nanoparticles function as a foam stability booster. As for nonionic surfactant/CNC stabilized foam, CNC boosts the foam stability.

By comparing the CTAB/CNC foam with nonionic surfactant/CNC foam, it is noted that CTAB/CNC foam is more stable than nonionic surfactant/CNC foam. This is probably because the electrostatic interaction of CTAB with CNC is stronger than the synergistic effect due to the use of nonionic surfactant and CNC, providing a more stable foam as a result.

3.3. Mobility of Foam Formed by CTAB-Coated CNC

First, we carry out a water-alternating-C_3H_8 gas (WAG) injection test with a water-to-gas ratio of 1:2 through a glass beadpack 20 °C, which serves as a baseline case for comparison with the following SAG tests. Then, the experiments are repeated by alternately injecting CTAB or CNC/CTAB and C_3H_8 at the same conditions. The concentrations of CNC and CTAB are kept as 0.300 w/v and 0.010 w/v, respectively. Figure 15 shows the measured differential pressures and mobility across the beadpack in these three tests. In the differential pressure profiles, the ascending segments represent the liquid-solution flooding stage, and the descending segments represent the C_3H_8 flooding stage. In comparison, in the mobility profile, the ascending segments represent the C_3H_8 flooding stage, and the descending segments mean the liquid-solution flooding stage. Figure 15 clearly indicates that both SAG (CNC/CTAB and CTAB alone) tests give higher pressure drops than the WAG test, indicating a substantial amount of foam within the beadpack is formed. As more and more foams are generated in the porous media, the pressure difference across beadpack will be increased until a steady state is reached [28]. In the presence of CNC nanoparticles, CNC/CTAB stabilized C_3H_8 foam gives a maximum pressure drop of around 1000 kPa at the steady state, which is much higher than that given by CTAB stabilized foam with a maximum pressure drop of 650 kPa. As foam flows through a porous media, its dynamic behavior is controlled by bubble snap-off, coalescence, movement, and trapping. If the foam is unstable, foam coalescence and regeneration occur when it propagates through porous media [29]. A higher pressure drop by CNC/CATB stabilized foam indicates an increased foam stability because CNC nanoparticles can be residing at the gas-quid interface, helping to prevent the foam from coalescence. As shown in Figure 15, CNC/CTAB stabilized foam shows a lower mobility than CTAB stabilized foam, and CTAB stabilized foam shows a lower mobility than the WAG case. This is because the addition of CNC nanoparticles forms more stable foam, which increases the apparent viscosity of the fluid and results in a higher mobility reduction effect. The key findings obtained from the mobility tests are consistent with those obtained in the former static foam-stability tests.

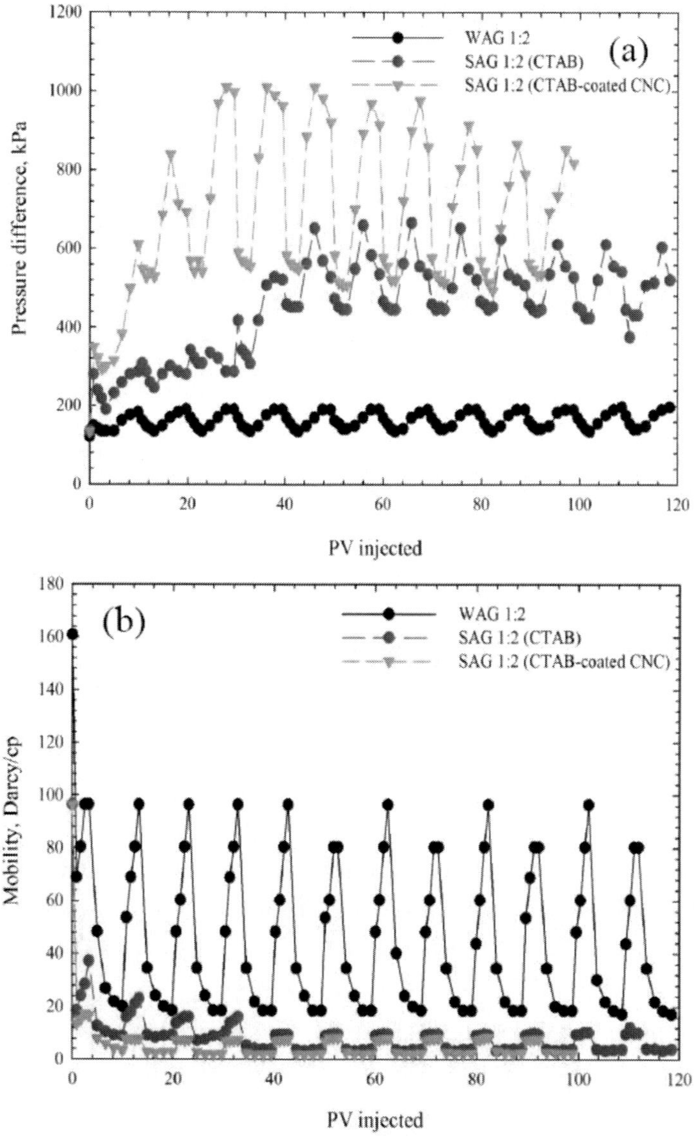

Figure 15. Comparison of pressure difference (a) and mobility (b) recorded in SAG tests (0.010 w/v CTAB, 0.3 w/v CNC, 20°C, SAG ratio of 1:2) and WAG tests (WAG ratio of 1:2).

3.4. Mobility of Foam Formed by Non-Ionic Surfactant and CNC Mixture

Based on the static foam stability tests for CNC in combination with nonionic surfactants in the previous section, Tween 20 and Triton X-100 (at their CMCs) are chosen to mix with CNC nanoparticles in the foam mobility tests. Four series of SAG flow experiments with the use of Tween 20, CNC/Tween 20, Triton X-100, and CNC/Triton X-100, respectively, are conducted at the same flow conditions as before. Figure 16 shows the test results on the pressure drop and mobility obtained for different cases. It can be found from Figure 16 that the SAG with CNC/Tween 20 exhibits a higher mobility reduction effect than the SAG with Tween 20, indicating a substantial amount of foam has been formed within the beadpack. The SAG with Tween 20 shows mobility profiles similar to WAG, implying little foam has been created by Tween 20 alone in the beadpack. During the test, we also do not have any foam collected at the outlet of the beadpack. Therefore, the addition of CNC into the Tween 20 solution is able to create foam in the beadpack, leading to a lower mobility than Tween 20 alone.

We conduct further tests to examine the mobility of foam formed by Triton X-100 alone and CNC/Triton X-100 mixture, respectively. Figure 17 shows the test results on the pressure drop and mobility obtained for different cases. The maximum steady-state pressure drops given by CNC/Triton X-100 mixture and Triton X-100 alone are about 750 kPa and 490 kPa, which are around 3.5 and 2.5 times of that by WAG, respectively. This proves that foam is formed in both SAG test. A lower mobility of foam formed by CNC/Triton X-100 is obtained than that formed by Triton X-100 alone, which indicates that the addition of CNC into nonionic surfactant makes the foam more stable in the porous media.

In summary, if a nonionic surfactant together with CNC is used as a foaming agent, the synergistic effect of CNC and a nonionic surfactant can lead to a lower mobility of foam in the porous media. Comparatively, when used together with CNC, CTAB displays a much better performance in terms

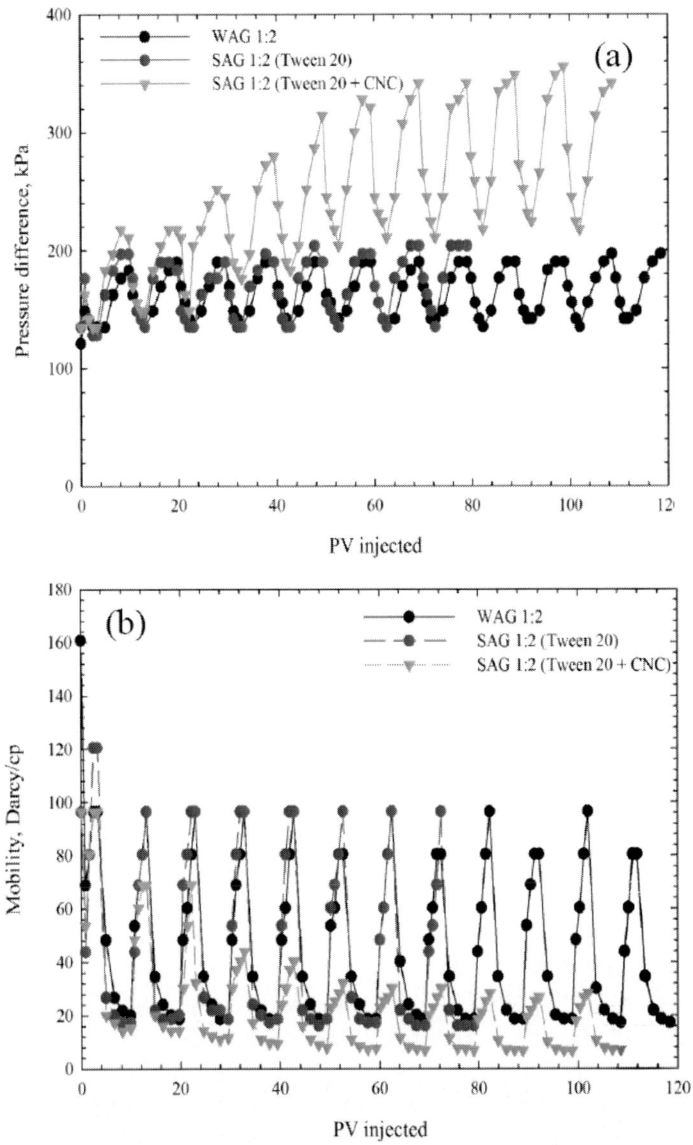

Figure 16. Comparison of pressure difference (a) and mobility (b) recorded in SAG tests (0.0071 w/v Tween 20, 0.3 w/v CNC, 20°C, SAG ratio of 1:2) and WAG tests (WAG ratio of 1:2).

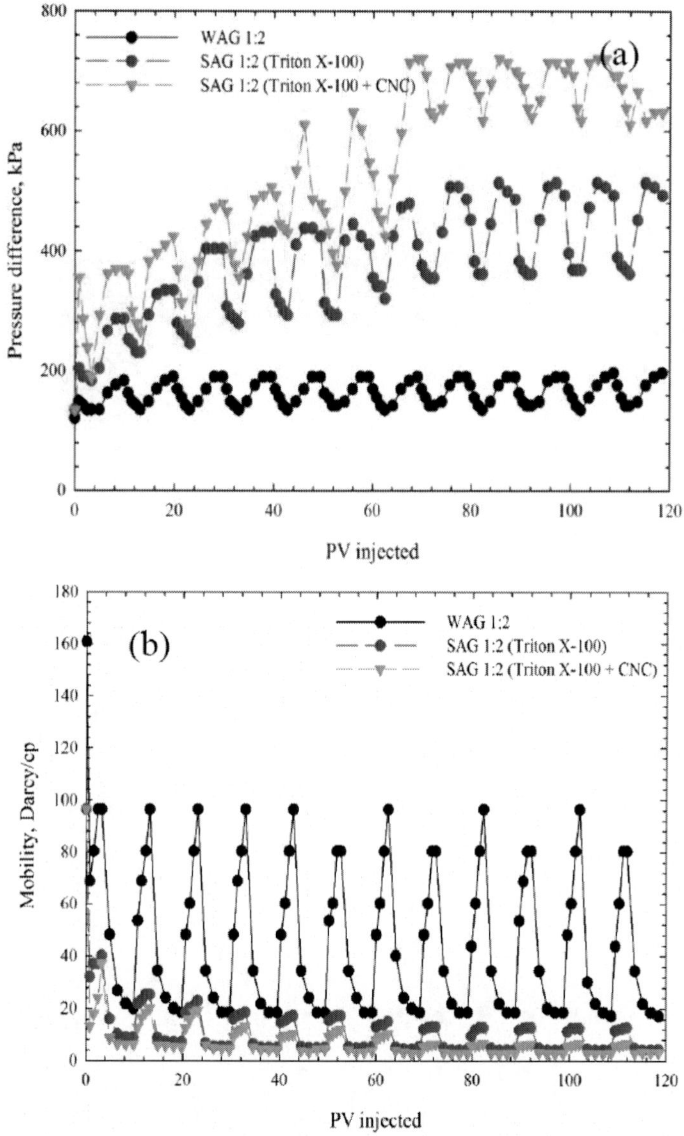

Figure 17. Comparison of pressure difference (a) and mobility (b) recorded in SAG tests (0.0194 w/v Triton X-100, 0.3 w/v CNC, 20°C, SAG ratio of 1:2) and WAG tests (WAG ratio of 1:2).

of mobility reduction of foam than other nonionic surfactants; this is probably attributed to a stronger electrostatic interaction existing between anionic CNC and cationic CTAB, preventing CNC from detaching from the bubble films. All these findings obtained from the mobility tests are consistent with those obtained from the static foam-stability tests.

3.5. Oil Recovery

In this section, further investigation on the ability of foam, formed by SAG, in recovering oil is carried out in n-$C_{16}H_{34}$ saturated beadpack under the same injection conditions as used in the previous tests. Firstly, a baseline test case is performed by alternately injecting water and C_3H_8 gas (WAG) into the oil-saturated beadpack. Then two series of experiments are repeated by either injecting CTAB/C_3H_8 or CTAB-coated CNC/C_3H_8 under the same conditions. The measured pressure drop and mobility across the beadpack as a function of PV injected are shown in Figure 18. It clearly indicates that, in the absence of CNC nanoparticle, the pressure drop is lower than the case where CNC nanoparticles are present. The CTAB-coated CNC case yields the maximum pressure difference as high as 348 kPa, 80 kPa higher than the CTAB-only case and around 2 times of the WAG baseline case. It is observed that a notable amount of resistance is provided by CTAB-coated CNC-foam despite the presence of oil in the beadpack. SAG (CTAB-coated CNC) yields the highest mobility reduction than the CTAB-only SAG and the WAG baseline case, indicating more stable foams formed in the porous media by CTAB-coated CNC. A lower mobility given by the CTAB-coated CNC indicates that the CNC nanoparticles can serve as an mobility control agent with the presence of oil when they are coated by CTAB.

Non-ionic surfactant Triton X-100 is selected to test the behavior of C_3H_8 foam formed by the combination of CNC with Triton X-100 in the oil-saturated beadpack. Two series of tests are conducted by alternately

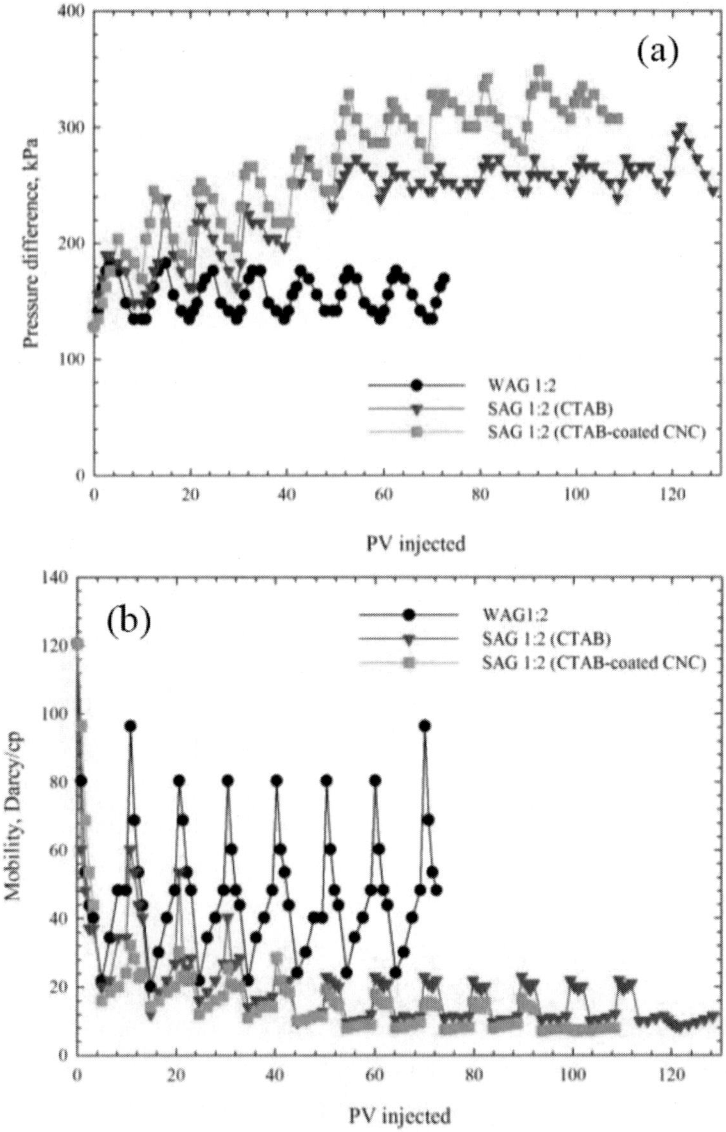

Figure 18. Comparison of pressure difference (a) and mobility (b) recorded in SAG tests (0.010 w/v CTAB, 0.3 w/v CNC, 20°C, SAG ratio of 1:2) and WAG tests (WAG ratio of 1:2) with the presence of oil.

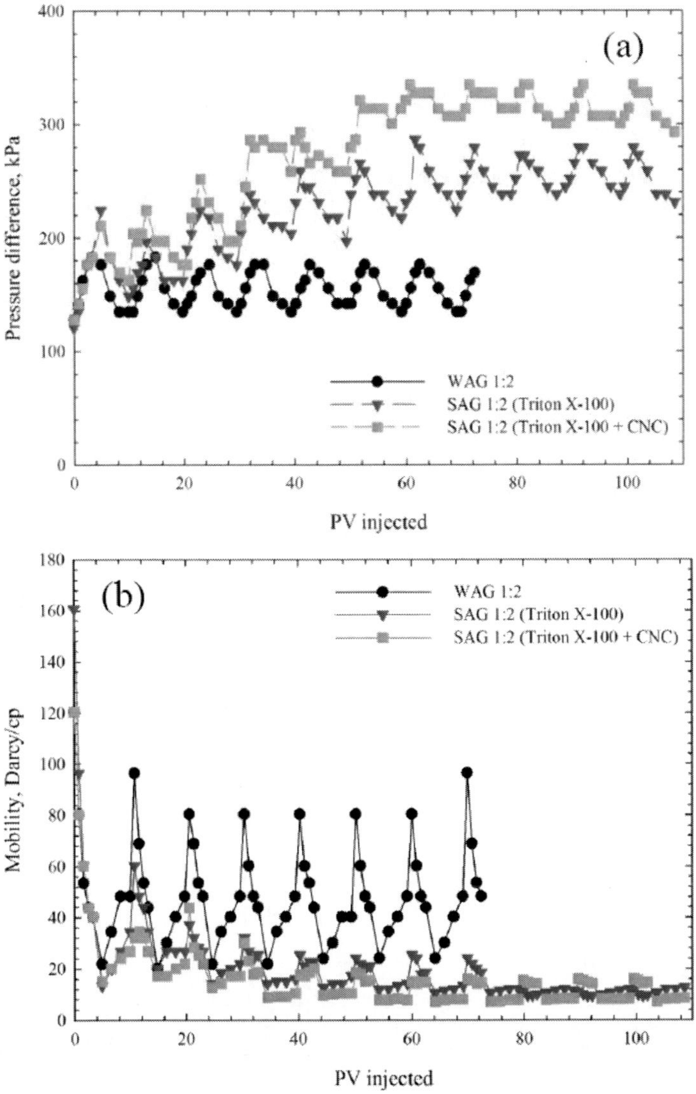

Figure 19. Comparison of pressure difference (a) and mobility (b) recorded in SAG tests (0.0194 w/v Triton X-100, 0.3 w/v CNC, 20°C, SAG ratio of 1:2) and WAG tests (WAG ratio of 1:2) with the presence of oil.

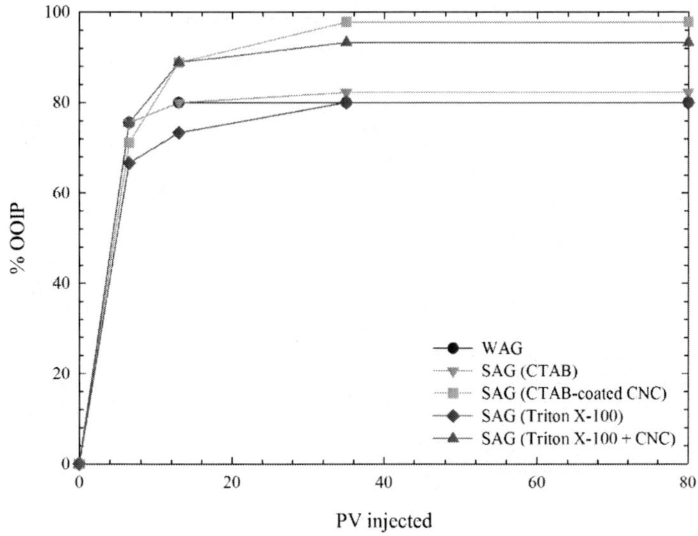

Figure 20. Comparison of total oil recovery profiles during WAG and different SAGs (CTAB, CTAB-coated CNC, Triton X-100, Triton X-100+CNC).

injecting CNC/Triton X-100 and C_3H_8 through the porous media under the same conditions as used in the previous tests. The measured results on pressure drop and mobility as a function of pore volume injected are shown in Figure 19. The maximum pressure drop measured in the CNC/Triton X-100 case is found to be about 334 kPa, 50 kPa greater than the Triton-only case and around 2 times of the WAG baseline case. The experimental case using CNC/Triton X-100 exhibits a lower mobility than the Triton-only case or the WAG case. The above discussion shows that CNC is able to serve as a mobility control agent when it is used in conjunction with the non-ionic surfactant Triton X-100.

During the SAG or WAG experiments, oil and chemical solution are collected from the outlet of the beadpack tube for further oil-recovery estimations. The cumulative oil recovery as a percentage of original oil in place (OOIP) is calculated for the WAG case and each SAG case. Figure 20 shows the oil recovery versus PV injected for the above 5 tests. Among all the cases, CTAB-coated CNC shows the best performance with an oil recovery of 97.8% OOIP, while water shows the lowest recovery with an oil recovery of 80% OOIP. The oil recovery given by the CTAB-coated CNC-stabilized foam (97.8%) is 15.6% OOIP more than that given by the CTAB-stabilized foam (82.2%). As for the test using Triton X-100/CNC dispersions, the addition of CNC leads to an additional oil recovery of 13.3% OOIP on top of the oil recovery given by Triton X-100. These results reveal that the foam formed by CNC nanoparticles can contribute to much more oil production than the foam generated by CTAB or Triton. In SAG, with the presence of oil, the nanoparticle-stabilized C_3H_8 foam reduces the residual oil saturation due to a mobility reduction mechanism.

CONCLUSION

In this study, the stability and mobility of foams generated by C_3H_8/surfactant/nanoparticle mixtures are evaluated experimentally by using a PVT and glassbead pack setups. The following conclusions can be obtained:

1) As for the CNC/CTAB stabilized foam, as the surface property of CNC changes from being hydrophilic to being relatively hydrophobic by adding cationic surfactant CTAB, more stable foams can be formed compared to the foam formed by CTAB alone.
2) As for the CNC/CTAB stabilized foam, its foam stability increases as salt concentration and temperature increase.
3) As for the CNC/CTAB stabilized foam, the presence of n-$C_{16}H_{34}$ (pseudo-oil) can only slightly decrease the foam stability.
4) If a surfactant is a foaming agent, the synergistic effect of CNC nanoparticle and non-ionic surfactant enhances the foam stability. In other words, a stable foam can be formed by introducing both foam forming agent (surfactant) and foaming booster (nanoparticle).
5) Foam stability of CNC/CTAB stabilized foam is stronger than that of CNC/non-ionic surfactant. This is probably because the electrostatic interaction of CTAB with CNC is stronger than the synergistic effect due to the use of non-ionic surfactant and CNC, leading to a more stable foam thereof.

6) If a non-ionic surfactant is used as a foaming agent in SAG, the synergistic effect of CNC and non-ionic surfactant enhances the reduction in fluid mobility, even under the influence of the hydrocarbon phase.
7) The foam formed by alternately injecting CTAB-coated CNC and C_3H_8 shows a higher mobility reduction effect than that formed by alternately injecting CNC/nonionic surfactant and C_3H_8.
8) In the SAG test [30] either CTAB-coated CNC or Triton X-100/CNC stabilized C_3H_8 foam) conducted with oil-saturated beadpack, the presence of CNC nanoparticles enhances the oil recovery due to the lowered foam mobility.

REFERENCES

[1] *BP Statistical Review of World Energy* 2014. May 2015. http://www.bp.com/en/global/corporate/about-bp/energy-economics/statistical-review-of-world-energy.html.

[2] Li, H., Zheng, S., Yang, D., 2013. Enhanced swelling effect and viscosity reduction of solvent/CO_2/heavy-Oil systems. *SPE J.*, 18 (4), 695-707.

[3] Upreti, S. R., Lohi, A., Kapadia, R. A., El-haj, R., 2007. Vapor extraction of heavy oil and bitumen: a review. *Energy Fuels* 21 (3), 1562-1574.

[4] Smith, D. H. *Surfactant-based mobility control: progress in miscible-flood enhanced oil recovery*. Washington DC: American Chemical Society, 1988.

[5] Grigg, R. B., Mikhalin, A. A. Effects of flow conditions and surfactant availability on adsorption. *Paper SPE-106205-MS presented at SPE Int. Symp. Oilfield Chem.,* Houston, Texas, U.S., February 28- March 2, 2007.

[6] Bond, D. C., Holbrook, O. C. U.S. Patent 2866507, 1958.

[7] Fred, A. N. The foam-drive process for increasing the recovery of oil, *Bureau of Mines Rept*. 1961, 5866.

[8] Bernard, G. G., Holm, L. W. *Method for recovering oil from subterranean formations.* U.S. Patent No. 3342256, 1967.

[9] Binks, B. P., 2002. Particles as surfactants: similarities and differences. Curr. Opin. *Colloid Interface Sci.* 7 (1), 21-41.

[10] Binks, B. P., Horozov, T. S. *Colloidal particles at liquid interfaces.* Cambridge Univ. Press 2006, 1-71.

[11] Ramsden W. Separation of solids in the surface layers of solutions and suspensions. *Proc. R. Soc. London*, 1903, 72, 156-164.

[12] Pickering S. U. Emulsions. *J. Chem. Soc.* 1907, 91, 2001–2021

[13] Alargova, R. G., Warhadpande, D. S., Paunov, V. N., Velev, O. D., 2004. Foam superstabilization by polymer microrods. *Langmuir* 20 (24), 10371-10374.

[14] Binks, B. P., Horozov, T. S., 2005. Aqueous foams stabilized solely by silica nanoparticles. *Angew. Chem.* 117 (24), 3788-3791.

[15] Espinoza, D. A., Caldelas, F. M., Johnston, K. P., Bryant, S. L., Huh, C. Nanoparticle-stabilized supercritical CO_2 foams for potential mobility control applications. *Paper SPE-129925-MS presented at the SPE Improved Oil Recovery Symposium*, Tulsa, Oklahoma, U.S., April 24-28, 2010.

[16] Yu, J., An, C., Mo, D., Liu, N., Lee, R. L. Foam mobility control for nanoparticle-stabilized supercritical CO_2 foam. *Paper SPE-153336-MS presented at SPE Improved Oil Recovery Symposium*, Tulsa, Oklahoma, U.S., April 14-18, 2012.

[17] Worthen, A., Bagaria, H., Chen, Y., Bryant, S. L., Huh, C., Johnston, K. P. Nanoparticle stabilized carbon dioxide in water foams for enhanced oil recovery. *Paper SPE-154285-MS presented at SPE Improved Oil Recovery Symposium*, Tulsa, Oklahoma, U.S., April 14-18, 2012.

[18] Zhang, S., Sun, D., Dong, X., Li, C., Xu, J., 2008. Aqueous foams stabilized with particles and nonionic surfactants. *Colloids Surf.* 324 (1-3), 1-8.

[19] Singh, R., Mohanty, K. K. Synergistic stabilization of foams by a mixture of nanoparticles and surfactants. *Paper SPE-169126-MS*

presented at SPE Improved Oil Recovery Symposium, Tulsa, Oklahoma, U.S., April 12-16, 2014.

[20] Cervantes Martinez, A., Rio, E., Delon, G., Saint-Jalmes, A., Langevin, D., Binks, B. P., 2008. On the origin of the remarkable stability of aqueous foams stabilized by nanoparticles: link with microscopic surface properties. *Soft Matter 4*, 1531-1535.

[21] Dickinson, E., Ettelaie, R., Kostakis, T., Murray, B. S., 2004. Factors controlling the formation and stability of air bubbles stabilized by partially hydrophobic silica nanoparticles. *Langmuir* 20 (20), 8517-8525.

[22] Deschenes, L. A., Zilaro, P., Muller, L. J., Fourkas, J. T., Mohanty, U., 1997. Quantitative measure of hydrophobicity: experiment and theory. *J. Phys. Chem. B* 101 (30), 5777-5779.

[23] Craig, V. S., Ninham, B. W., Pashley, R. M., 1993. The effect of electrolytes on bubble coalescence in water. *J. Phys. Chem.* 97 (39), 10192-10197.

[24] Wang, L., Yoon, R. H., 2006. Role of hydrophobic force in the thinning of foam films containing a nonionic surfactant. *Colloids Surf. A: Physicochem. Eng. Aspects* 282-283, 84-91.

[25] Schramm, L. L. Foam sensitivity to crude oil in porous media. In *Foams: fundamentals and applications in the petroleum industry*, Schramm, L. L., Ed., American Chemical Society: Washington, DC, 1994, Chapter 4, pp. 165-197.

[26] Pugh, P. J., 1996. Foaming, foam films, antifoaming and defoaming. *Adv. Colloid Interface Sci.* 64, 67-142.

[27] Wang, C., Li, H., 2016. Stability and mobility of foam generated by gas-solvent/surfactant mixtures under reservoir conditions. *J. Nat. Sci. Eng. Sci.* 34, 366-375.

[28] Yang, S. H., Reed, R. L. Mobility control using CO_2 foams. *Paper SPE-19689-MS presented at the SPE Annual Technical Conference and Exhibition*, San Antonio, Texas, October 8-11, 1989.

[29] Holm, L. W., 1968. The mechanism of gas and liquid flow through porous media in the presence of foam. *SPE J.* 8 (4), 359-369.

[30] Belhaij, A., AlQuraishi, A., Al-Mahdy, O. Foamability and foam stability of several surfactants solutions: the role of screening and flooding. *Paper SPE 172185 presented at the SPE Saudi Arabia Section Technical Symposium and Exhibition*, Al-Khobar, Saudi Arabia, April 21-24, 2014.

List of Contributors

Aman Ullah
Department of Agricultural, Food and Nutritional Science
University of Alberta, Edmonton, Alberta, Canada.

Abdullahil Kafy
CRC for Nanocellulose Future Composite, Inha University
Incheon city, Republic of Korea.

Alfred C. W. Leung
Aquatic and Crop Resource Development Research Center
National Research Council of Canada, Montreal, Quebec, Canada.

Behzad Ahvazi
InnoTech Alberta Inc., Edmonton, Alberta, Canada.
Department of Mechanical Engineering, Faculty of Engineering
University of Alberta, Edmonton, Alberta, Canada.

Chao Wang
Department of Civil and Environmental Engineering
University of Alberta, Edmonton, Alberta, Canada.

Christie Sayes
Department of Environmental Science
Baylor University, Waco, TX, USA.

Christophe Danumah
InnoTech Alberta Inc.
Edmonton, Alberta, Canada.

Dinh-Chi Pham
Global Engineering and Material Inc.
1 Airport Place, Princeton, NJ 08540, USA.

Douglas M. Fox
Department of Chemistry
American University, Washington, DC, USA.

Edmond Lam
Aquatic and Crop Resource Development Research Center
National Research Council of Canada, Montreal, Quebec, Canada.

Huazhou Andy Li
Department of Civil and Environmental Enginering
University of Alberta, Edmonton, Alberta, Canada.

Hyun Chan Kim
CRC for Nanocellulose Future Composite, Inha University
Incheon city, Republic of Korea.

J. I. Cruz
Department of Chemistry
American University, Washington, DC, USA.

Jaehwan Kim
CRC for Nanocellulose Future Composite, Inha University
Incheon city, Republic of Korea.

Jeffrey W. Gilman
Materials Science and Engineering Division,
Materials Measurement Laboratory
National Institute of Standards and Technology, Gaithersburg, MD, USA.

Jeremiah Woodcock
Materials Science and Engineering Division,
Materials Measurement Laboratory
National Institute of Standards and Technology, Gaithersburg, MD, USA.

Jung Woong Kim
CRC for Nanocellulose Future Composite, Inha University
Incheon city, Republic of Korea.

Katrina Wakeman
Department of Chemistry
American University, Washington, DC, USA.

Le Van Hai
Dept. of Pulp and Paper Technology
Phutho College of Industry and Trade, Phutho, Vietnam.

Lindong Zhai
CRC for Nanocellulose Future Composite, Inha University
Incheon city, Republic of Korea.

Muhammad Arshad
Department of Agriclutural, Food and Nutritional Science
University of Alberta, Edmonton, Alberta, Canada.

Muhammad Safder
Department of Agricultural, Food and Nutritional Science
University of Alberta, Edmonton, Alberta, Canada.

Muhammad Zubair
Department of Agricultural, Food and Nutritional Science
University of Alberta, Edmonton, Alberta, Canada.

Rehan Ali Pradhan
Department of Agricultural, Food and Nutritional Science
University of Alberta, Edmonton, Alberta, Canada.

Ruth M. Muthoka
CRC for Nanocellulose Future Composite, Inha University
Incheon city, Republic of Korea.

Tess Ravick
Department of Chemistry
American University, Washington, DC, USA.

Whirang Cho
Department of Chemistry
American University, Washington, DC, USA.

About the Editor

Tri-Dung Ngo, PhD, Research Scientist, Adj. Prof.

Biomass Conversion and Processing

InnoTech Alberta, Edmonton, Alberta, Canada

Department of Civil & Environmental Engineering

Faculty of Engineering, University of Alberta, Edmonton, Alberta, Canada.

Dr. Ngo received his Bachelor and Master in Chemical Engineering from Polytechnic University (Ho Chi Minh University of Technology), Vietnam. In 1997, he held a position as lecturer and researcher in Department of Chemical and Engineering at Polytechnic University. In 2001, he joined Department of Materials Engineering at Polytechnic University as lecturer and researcher. He obtained his Ph.D. in Mechanical Engineering from Concordia University, Montreal, Canada in 2007. From 2003 to 2007, he worked as invited researcher at Industrial Materials Institute (IMI) of National Research Council of Canada (NRC). After completing his Natural Sciences and Engineering Research Council of Canada (NSERC) post-doctoral fellowship in 2008, he joined IMI of NRC as researcher. During his tenure, he had performed innovative R&D activities on polymer, composite, nanocomposite, biopolymer, biocomposite, biofoam, including synthesis, modification, formulation, processing and performance of biomaterials from biomass for industrial applications.

Dr. Ngo is currently working at InnoTech Alberta (Formerly Alberta Research Council (1921-2010) and Alberta Innovates Technology Futures (2010-2016)) as research scientist for Biomass Conversion and Processing Technologies, a position he has appointed since late 2014. His research focuses on biomass processing and development of novel, eco-friendly applications using renewable and sustainable biomaterials including but are not limited to polymer, biocomposite, nanocomposite, celluloses, nanocelluloses, and lignin etc. Dr. Ngo is also an adjunct professor in the Department of Civil & Environmental Engineering at University of Alberta, Edmonton, Canada.

INDEX

#

A

B

C

D

E

F

G

H

I

K

L

M

N

Q

R

S

T

U

V

W

X

Z

Nanotechnology: Principles, Applications and Ethical Considerations

Editors: Paresh Chandra Ray (Jackson State University, Jackson, MS, USA)

Series: Nanotechnology Science and Technology

Book Description: The volume explores the emerging science of nanotechnology which deals with the understanding of the fundamental physics, chemistry, biology, material science and technology of nanometer scale objects, which has become a central pillar for the next generation medical challenges such as developing tiny nanodevices, as well as for food technology.

Hardcover ISBN: 978-1-53613-889-4
Retail Price: $195

Fiber-Based Optical Trapping and Manipulation

Authors: Baojun Li and Hongbao Xin (Institute of Nanophotonics, Jinan University, Guangzhou, China)

Series: Nanotechnology Science and Technology

Book Description: Since its first report in 1970 by A. Ashkin, optical trapping and manipulation has been widely used in the interdisciplines of micro- and nano-photonics, biophotonics, biomedicine, etc.

Hardcover ISBN: 978-1-63485-998-1
Retail Price: $160

First-Principle vs. Experimental Design of Nanomaterials

Authors: Omar Mounkachi, Abdelilah Benyoussef, and Mohamed Hamedoun

Series: Nanotechnology Science and Technology

Book Description: The purpose of this book is to propose some ideas to answer the most important question in the design of nanomaterials (OD,1D and 2D) for nanotechnology application, namely, nanomaterials for spintronic application, nanomaterials for solar energy technologies application, magnetic refrigeration applications, switchable materials application and nanomedicine applications.

Softcover ISBN: 978-1-53613-984-6
Retail Price: $95

Carboxymethyl Cellulose. Volume II: Pharmaceutical and Industrial Applications

Editor: Md. Ibrahim H. Mondal

Series: Biochemistry Research Trends

Book Description: Highlights of this book include synthesis and formation mechanisms of CMC, structure property relationships, homogeneous and heterogeneous phase modification to create common and functional products, analysis and characterization, absorption and swelling mechanisms, stimuli-responsive and super-absorbency nature, and a wide range of multi-dimensional applications, including human consumption and uses.

Hardcover ISBN: 978-1-53614-751-3
Retail Price: $230